AGRICULTURE AND FERTILIZERS

Other Related Books

Organic Agriculture and Climate Change Mitigation	*URS Negi*
Drainage Engineering	*James N. Luthin*
Soil Conservation Technical Handbook	*Denise F. Church*
Fertilizers and Their Use	*FAO*
Physiological Genetics of Agricultural Crops	*Andor Balint*
Soil Conservation	*Normal Hudson*
Guide to Laboratory Establishment for Plant Nutrient Analysis	*M.R. Motsara*
Plant Physiology in Relation to Horticulture (2nd Ed.)	*Bleasdale, J.K.A.*
Horticultural Engineering Technology Field Machinery	*Balls, R.C.*
Economics in Horticulture	*M.J. Sargent*
Application of Botany in Horticulture	*Andre Genin*
Chemistry of the Soil (2nd Ed.)	*Firman E. Bear*
Soil Chemical Analysis (2nd Ed.)	*M.L. Jackson*
Pest and Vector Management in the Tropics	*Anthony Youdeowei*
The Soil-Plant System in Relation to Inorganic Nutrition	*Maurice Fried*

AGRICULTURE AND FERTILIZERS

FERTILIZERS IN PERSPECTIVE

- Their role in feeding the world
- Environmental challenges
- Are there altenatives?

By Oluf Chr. Bøckman
Ola Kaarstad
Ole H. Lie
Ian Richards

United Book Prints

1st Reprint in India, 2015

Published by:
United Book Prints
4806/24, Ansari Road,
Daryaganj,
New Delhi - 110 002
Tel.: 011-41511055
E-mail: unitedbookprints@gmail.com

Print : 2021

Print: 2015

ISBN: 978-93-83692-10-1

Printed in India

TABLE OF CONTENTS

PREFACE

A growing number of people, particularly in the industrialized nations, are alarmed over threats to nature, to the environment and ultimately to man as a result of prevailing agricultural, industrial and social practices.

The threats comprise pollution, destruction of habitats, depletion of mineral and biological resources, possible changes in weather patterns caused by the burning of fossil fuels (the greenhouse effect), depletion of the ozone layer and possible links between these factors and cancer and other diseases.

There are good reasons to be concerned about the environment on our planet. The extraordinary growth in world population during the last centuries - a growth which continues unabated - and the rise in consumption per capita are putting a great strain on resources and the environment. In most instances the effects are local, but there are signs that the global environment is changing due to the actions of man.

Modern agriculture is seen by many as a major contributor to some of the present environmental ills. Agriculture is considered to be a prime source of pollution - eutrophication of fresh and marine waters, increasing nitrate concentrations in ground and surface waters, and pesticide residues in soil, water and food.

It is true that agriculture - like the rest of society - has been less environmentally conscious than could be desired, and appropriate regulations and changes of practice are required.

Environmental protection is on the agenda throughout the world. We have all - as individuals, as members of groups and organizations, as citizens of nations and ultimately as inhabitants of the planet earth - a responsibility to partake in the debate on the basis of our particular knowledge on specific topics and to contribute to the proper remedial actions. Norsk Hydro is no exception.

The issues relating to agriculture and the environment are very complex. There exists a large body of scientific knowledge in this field. The public debate does not always reflect these insights. It is of the utmost importance to use what is known, and to seek new knowledge where needed. Regulatory actions based on insufficient insights can lead to situations where the cures are worse than the ills.

Norsk Hydro's Agricultural Group, as a producer of 12 million tonnes of fertilizer annually, has an obvious interest in the future of agriculture. We have also a large knowledge base about agriculture and fertilizer use, accumulated over the company's 85 years of experience as a fertilizer producer in the international market.

The purpose of this book is to provide factual information about the environmental aspects of modern agriculture, with an emphasis on the issues relating to fertilizer use. The information is drawn from the scientific literature and the text has been reviewed by scientists at universities and independent research institutes.

It is our firm conviction that fertilizer, when properly used, is an environmentally benign product which is indispensable in agriculture's task of feeding the world's population. We believe the material presented here substantiates this conviction.

That is not to say that modern agriculture using fertilizer is able to feed an unlimited population. It is evident that population growth must be controlled and eventually stopped. Neither is it to say that modern agricultural practices, including fertilizer use, cannot be improved and be made more environmentally sound. But, we are concerned that some consider fertilizer use *per se* to be the root of agriculture's environmental problems. Left unchallenged, such beliefs may do agriculture, society - *and* the environment - irreparable harm.

There is a long tradition in Norsk Hydro, and in the fertilizer industry, of providing the farming community with information about the use of fertilizers. However, increased emphasis must be put on the environmental aspects of fertilizer use, both in the information, recommendations and service to the farmer and in development of new fertilizer products. Norsk Hydro accepts this challenge.

The appraisal of agronomic practices is no longer the exclusive domain of the farming community. We are therefore addressing this book to everybody who cares for our common environment.

It is our hope that this book will contribute to a better understanding of the complexities of agriculture, and will be of value in future discussions of regulations and changes of practices.

Oslo, April, 1990

Trygve Refvem
President, Agricultural Group
Norsk Hydro a.s

INTRODUCTION

The intention of this book is to give a survey of environmental issues relating to agriculture and fertilizer use, with an outline of present conflicts of opinion and the status of knowledge. Pointers are given to books, reviews and papers where further details can be found.

Fertilizers, what they are and how they are used, are described in Chapter 1.

The main challenges to agriculture - to feed the present and future generations while caring for our common environment - are outlined in Chapter 2.

Some knowledge of the past is useful in understanding the present. A description of the developments in European agriculture after the Second World War is given in Chapter 3.

The issues relating to fertilizer use in conventional agriculture - or current agriculture, the term we have elected to use in this book - are discussed in more detail in Chapter 4.

Various alternatives to current agricultural practices are being proposed. Alternative agriculture should be examined in the same critical manner as current agriculture. This is the subject of Chapter 5.

Our own reflections after having completed this book are summarized in "Final Comments" at the end of the text.

The book is intended for anybody with an interest in agriculture and the environment. Some familiarity with science and environmental topics would facilitate the reading. Considerations of size and readability have limited the amount of details that can be presented on each issue.

It is necessary to use some scientific terms for brevity and precision. These terms and units of measurements have been defined in the Appendix. Reference list and Subject index can be found at the end of the book.

* * * * *

In many respects, this book is the result of a collective effort within the Agricultural Group of Norsk Hydro. Colleagues in the various companies within the Group - Holger Asfeldt (Denmark), Göte Bertilsson (Sweden), Nicolas Ferenczi (France), Dies Koole (Belgium), Hermann Kuhlmann

(Germany) and Trond Gulbrandsen, Gunnar Kongshaug and Bernt Stenrød (Norway) - have all given significant contributions. Many others have given valuable comments and suggestions. Annelise Lehmann has cheerfully handled the manuscripts and the many revisions. Ole Johan Sagafos and Gry Andersson have been very helpful in the process of converting the manuscript into a book.

We thank the copyright-holders for permission to reproduce their material in photographs, figures and tables. Experts from scientific communities in Europe and the USA have generously commented on the drafts of the various chapters. Their critical and constructive comments are gratefully acknowledged. The sometimes conflicting views expressed in these comments reflect the lively scientific debate and development which are characteristic of many aspects of the subjects.

Oslo, April, 1990

Oluf Chr. Bøckman
Ola Kaarstad
Ole H. Lie
Ian Richards *

* Presently with Levington Agriculture Ltd., UK.

THE BASIS FOR FOOD PRODUCTION: THE PLANT NUTRIENTS

Food and fertilizers

We all, 5 000 millions of us, depend on plants for our food, and plants depend on mineral nutrients for their growth and development.

13 elements derived from the soil are indispensible for all plant growth. They are called plant nutrients. An additional 4 or 5 elements are beneficial for proper development of some plants (Barber 1984, Marschner 1986).

Fertilizers are plant nutrients.

Plant nutrients

Plants form their complex organic matter from water and nutrients from the soil, carbon dioxide from the air and the energy from sunlight.

Plants use six of the nutrients in relatively large amounts: nitrogen, phosphorus, potassium, sulphur, calcium and magnesium. These are called "major nutrients". They are constituents of many plant components such as proteins, nucleic acids and chlorophyll, and are essential for processes such as energy transfer, maintenance of internal pressure and enzyme function.

The other nutrients are required in small or trace quantities and are referred to as "micronutrients" or "trace elements". They have a variety of essential functions in plant metabolism. The metals are constituents of enzymes.

Table 1.1 lists the plant nutrients, their main chemical form of uptake from the soil, and typical amounts removed from the land with a 5t/ha harvest of wheat grain.

Table 1.1 PLANT NUTRIENTS

Nutrients and their chemical symbols	Principal chemical form taken up by plants	Nutrient removal with a wheat harvest (5t/ha, 20 % moisture)
Macronutrients		kg/ha
Nitrogen (N)	NH_4^+, NO_3^-	105
Phosphorus (P)	$H_2PO_4^-$	18
Potassium (K)	K^+	15
Sulphur (S)	SO_4^{2-}	8
Magnesium (Mg)	Mg^{2+}	6
Calcium (Ca)	Ca^{2+}	2
Micronutrients		
Chlorine (Cl)	Cl^-	3
Iron (Fe)	Fe^{2+}	0.2
Manganese (Mn)	Mn^{2+}	0.2
Zinc (Zn)	Zn^{2+}	0.2
Copper (Cu)	Cu^{2+}	0.03
Boron (B)	H_3BO_3	0.02
Molybdenum (Mo)	MoO_4^{2-}	–

Source: IAEA (1984).

Other elements required by some plants:
sodium, cobalt, vanadium, silicon, nickel (Loue 1986).

Animals have a similar but slightly different list of essential elements:
man requires iodine and selenium but not boron.

There are two ways of expressing plant nutrient contents. In both, nitrogen content is expressed as the element nitrogen (N). Phosphorus and potassium are either given as the content of the oxides (P_2O_5 and K_2O) or as the elements (P and K). The latter system is in increasing use and is used here unless otherwise stated.

For optimum plant growth, nutrients must be available for plants:

- in solution in the soil water
- in appropriate and balanced amounts
- at the right time.

When deficiencies or gross imbalances occur, plant growth and development suffer. Figures 1.1 and 1.2 show a series of plants with characteristic signs of nutrient deficiencies.

Plants are supplied with nutrients mainly from:

- release of nutrients from soil reserves
- decomposing plant residues (roots, straw, etc.)
- organic manures
- mineral fertilizers
- biological nitrogen fixation
- aerial deposition.

Nutrients removed from the soil must be replenished, otherwise the soil becomes exhausted and crops will suffer and eventually fail.

Soils contain reserves of nutrients, e.g., the topsoil content of nitrogen ranges from some 3 to 20 t/ha. However, these reserves are mostly in forms unavailable to plants; only a minor portion is released each year through biological activity or chemical processes. Plants can only take up nutrients as water soluble compounds.

When the nutrient supply is insufficient for crop needs, additional nutrients can be supplied in fertilizers to make up the difference. Mineral fertilizers are not substances foreign to nature: they contain normal plant constituents. Their use is based on about 150 years of experience.

Figure 1.1 NUTRIENT DEFICIENCIES IN BARLEY

Plants with normal, balanced nutrient supply to the left.

Lack of nitrogen.

Lack of sulphur.

Lack of magnesium.

Yellow stripes on barley leaves, a typical symptom of magnesium deficiency.

Demonstration pots at the Department of Soil Sciences, Agricultural University of Norway. Photos: T. Knudsen, Norsk Hydro.

Figure 1.2 NUTRIENT DEFICIENCIES IN TOMATO PLANTS

Plants with normal balanced nutrient supply to the right.

Potassium deficiency.

Nitrogen deficiency.

Severe phosphorus deficiency.

Demonstration pots at the Department of Soil Sciences, Agricultural University of Norway. Photos: T. Knudsen, Norsk Hydro.

Figure 1.3 FERTILIZER PRODUCTION ROUTES

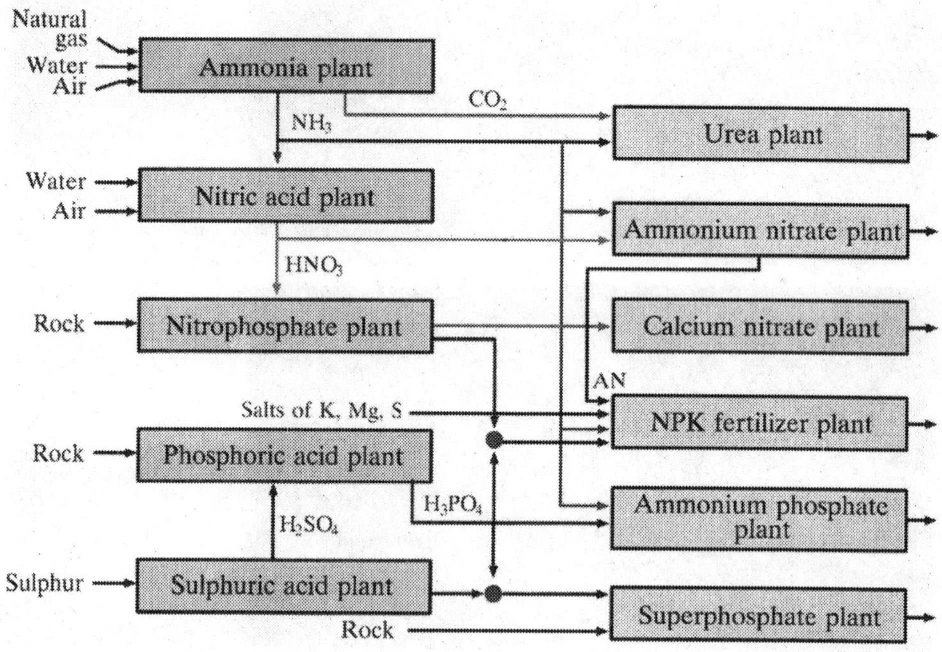

Fertilizers

Fertilizer manufacture

A simplified diagram for fertilizer production is given in Figure 1.3.
Fertilizer nitrogen originates from ammonia made by industrial synthesis from the elements hydrogen and nitrogen. The importance of this process was recognized by the award of Nobel prizes in chemistry to Fritz Haber (1918) and to Carl Bosch (1931). Hydrogen is mostly made through the reaction between water and fossil fuel; natural gas is the preferred raw material. Nitrogen is supplied by air. The overall reaction is:

air + natural gas + water -> ammonia + carbon dioxide

$$O_2 + N_2 + CH_4 + H_2O \rightarrow NH_3 + CO_2$$

Carbon dioxide from this conversion is combined with the ammonia to form urea, a major fertilizer product (urea is also the main nitrogen compound in urine from mammals). Oxidation of ammonia gives nitric acid. This, combined with ammonia, gives ammonium nitrate, another major fertilizer product, and a key component of complex fertilizers.

The raw material for phosphate fertilizers is rock phosphate. This mineral varies in properties depending on origin. Most rock phosphates are almost insoluble in soil water and thus unavailable to plants. Phosphate fertilizers are made by dissolving the rock in sulphuric and/or phosphoric acid to give superphosphates (straight P-fertilizers). More commonly, the rock is dissolved in sulphuric or nitric acid and calcium is removed as calcium sulphate (gypsum, a process waste) or as calcium nitrate (a fertilizer material). The remaining acid is neutralized with ammonia to ammonium phosphates. These are fertilizer materials and also key components in complex fertilizers.

Potassium chloride from natural deposits of salts or brine is the main source of potassium. Potassium chloride can be used alone or in combination with other fertilizers. Potassium chloride can be converted to potassium sulphate for use in chloride free fertilizers, required by some crops. Magnesium is added to a variety of fertilizer products as natural carbonates or processed sulphates. Other nutrients are added as required.

An important part of fertilizer manufacture is particulation, e.g., preparing the finished product as well-rounded particles that can be stored and transported without damage to the product and spread evenly in the field. Even spreading permits controlled application of nutrients to the plants and helps to maintain fields in a uniform state of fertility.

Prilled mineral fertilizer (NPK).
Photo: T. Knudsen, Norsk Hydro.

The basic principles of fertilizer manufacture are simple. In practice there are many technical difficulties that require special knowledge and experience. For safety, economy and efficiency, fertilizer plants are large and have a complicated structure. A new factory for complex fertilizer (with nitrogen, phosphate and potassium) starting from natural gas may make about 2500 t/d of product and cost some USD 600 million to build.

Fertilizer application

Most fertilizers are applied by surface spreading. In arable crops sowing and fertilization are sometimes combined in one operation, with the fertilizer placed near the seeds. Later applications of nitrogen are spread on the surface. Injection of liquid ammonia into the soil using special equipment and the spreading of fertilizers dissolved in water are also used in some regions.

Plants take up most of their nutrients from the soil solution through the roots, but they can also take up some nutrients sprayed on the leaves. This is the usual application method for correcting deficiencies of micronutrients.

Surface application of mineral fertilizer by a centrifugal spreader.
Photo: Norsk Hydro.

Fertilizers should be used according to fertilizer recommendations published by governmental and agricultural agencies and by fertilizer producers. Increasingly, fertilizer plans are made for each field. Crop requirements, nutrient supply from soils as determined by soil analysis, residues from past cropping, manure application and local soil and climatic conditions are all important in estimating the fertilizer rate. Application timing is also important. Needs vary with the stage of plant development. Too little fertilizer reduces crop yields, too much is wasteful and may result in environmental problems.

References:
Barber (1984), Cooke (1982), FAO (1984), Marschner (1986), Mengel and Kirkby (1987), Tisdale et al. (1985), Wild (1988).

Manure

Organic manure can be of plant or animal origin or a mixture of both. The largest quantities derive from the dung and urine of farm animals.

Animal manure

Depending on the method used for collection and storage, manure can occur in various forms: dry, wet (urine), slurry (mixed dry and wet) or as a compost. Nutrient content depends on the species of animal, type of feed and method of storage. Manure is a source of organic matter and contributes to soil structure and humus content.

Sewage sludge is used in some areas as a manure though it often has the disadvantage of a high heavy metal content (see Section 4.6).

The water content of manure produced on farms tends to be high and the nutrient content low and variable. Table 1.2 shows average nutrient contents for some manures.

Table 1.2 MANURE PRODUCTION AND NUTRIENT CONTENT

Nutrients kg/t slurry/manure (with the dry matter content listed)

Source	Manure t/year	% dry matter	Total Nitrogen N	Soluble Nitrogen N	Phosphorus P	Potassium K
1 Dairy cow (Slurry)	23.0	8.5	4.7	2.7	0.6	4.4
10 Pigs (Slurry)	21.0	6.6	6.3	4.4	1.5	2.9
100 Layer hens						
Slurry	10.0	13.8	5.4	3.5	2.6	2.5
Manure	2.3	71.0	21.7	5.9	15.8	16.3
1000 Chickens (Manure)	1.1	57.4	21.5	6.5	14.3	17.0

Source: Danish Agricultural Information Office (1989).

Part of the nutrients in manure is in water-soluble form and immediately available to crops. The rest is in insoluble organic matter and must be decomposed (mineralised) by micro-organisms before becoming available. The rate of this process depends on many factors so it is difficult to predict the amount and timing of nutrient release following application of organic manure.

Guano, accumulated droppings of birds, seals and other wild animals, is a traditional fertilizer. Peru is the major supplier, but the production is only about 25 thousand tonnes per year. This is quite insignificant as a nutrient supply on a world scale.

Manure is a resource that should be utilized fully where available. But the application of manure only returns nutrients to soil. It does not compensate for nutrient losses and exports from the farm unless animal feed is brought in from outside. Farmyard manure was traditionally surface applied on arable land and cultivated into the soil. Liquid manure is mostly surface

Field manuring in India. Dung and other forms of manure are valuable nutrient sources.
Photo: Jarle Rognaldsen, Oslo.

spread on grassland. Injecting liquid manure reduces ammonia losses and this practice is increasing. Manure handling requires labour and special equipment. Some regions (e.g., Southern Netherlands) have so many animals fed on imported feed that manure is produced in excess of local needs. Manure is bulky and long distance transport uneconomic. For this reason, drying processes are being developed that make a product which typically contains about 4 per cent nitrogen, 2 per cent phosphorus and 5 per cent potassium. However, processing costs are considerable.

References: Kofoed (1985), van der Meer et al. (1987).

Green manure

When fresh plant material is added directly to the soil without composting or passing through animals, it is termed "green manure". Use of green manure helps to prevent erosion and conserves nutrients but does not add nutrients except when legumes are used as a source for nitrogen.

The growing of plants (e.g., grass or rape) in the autumn in order to take up nitrate that would otherwise be leached is becoming more common. Such plant material should be ploughed down as green manure in spring. However, this is difficult on heavy soils where autumn ploughing is necessary.

Biological nitrogen fixation

Some bacteria are able to convert nitrogen gas from the air to ammonia for use as their nitrogen source. This biological nitrogen fixation is an energy demanding process, and requires about 10 kg of sugar equivalent as a bacterial energy source for each kg of nitrogen fixed. Biological nitrogen fixation by free-living bacteria in the soil provides a modest contribution to soil nitrogen supply: less than 10 kg per ha per year. Plants cannot fix nitrogen, but some can accept and control root infection with nitrogen fixing bacteria: these supply the plant with nitrogen (ammonia) while the plant provides the bacteria with energy in the form of organic acids. Notable examples of such mutually beneficial relationships (symbiosis) are found in the legumes, e.g., clover, alfalfa, peas, beans. Another example is the small water fern Azolla that harbours nitrogen fixing cyanobacteria. This fern has long been grown in rice paddies in South East Asia as a nitrogen source and as a way of shadowing out weeds. Some non-leguminous trees and plants (e.g., alder, sugar cane) also harbour nitrogen fixing bacteria.

The amount of nitrogen fixed in symbiotic nitrogen fixation varies greatly with the crop species; it ranges from some tens to some hundreds of kg nitrogen per ha per year. Ample nitrogen supply from other sources suppresses biological nitrogen fixation. The legume plant residues must decompose before the nitrogen can be used by other crops. There is little direct transfer of nitrogen between living legumes and other plants, even when they are neighbours.

Generally, fertilizers have supplanted the use of plants grown mainly for their nitrogen fixing properties. One reason is that fertilizers allow larger and more controlled nitrogen input and higher yields. The advantages and disadvantages of legumes as a nitrogen source in agriculture are described in more detail in Section 4.3 and 5.3.

References: Dixon and Wheeler (1986), Postgate (1982).

Root of a bean plant, with nodules (growths) harbouring nitrogen fixing bacteria.

Photo: T. Knudsen, Norsk Hydro.

CHAPTER 2

THE GLOBAL CHALLENGE: TO FEED THE PEOPLE

2.1 Population growth and food requirements

United Nations (UN) median projections indicate that world population will increase from 4 800 millions in 1985 to 6 100 millions in 2000, Figure 2.1.1.

Figure 2.1.1 POPULATION GROWTH, 1950 - 2025

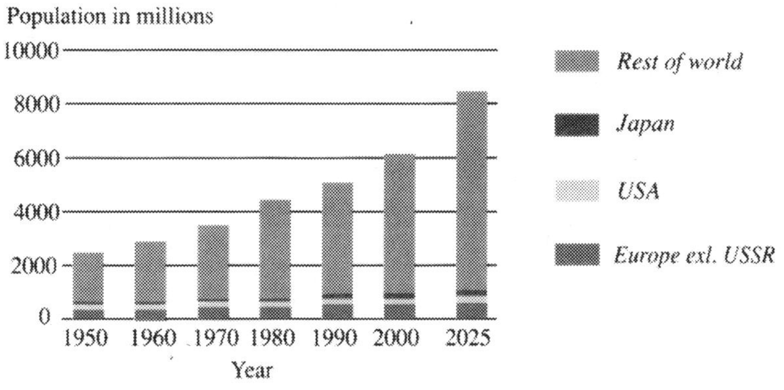

Source: UN (1989).

A population of 10 000 millions is expected to be reached later in the next century. More than 90 per cent of the increase will occur in developing countries. The increase in population along with the anticipated increase in per capita income will cause the demand for food and other agricultural products to expand at an annual rate of 3.1 per cent.

Present food demands can be met, but increased future needs will put agriculture and its resources under severe strain especially in some of the developing countries. Increasing instances of regional shortage are probable.

Total food production increased on average at an annual rate of 3.2 per cent in developing countries and by 2.0 per cent in developed countries between 1961 and 1985. UN Food and Agriculture Organization (FAO) estimates indicate that about one-third of the increase was due to increases in harvested area and two-thirds to increases in crop yields.

Average wheat yields in developing countries increased by 70 per cent and rice by 41 per cent between 1969 and 1985. New crop varieties with improved yield potential which can be protected by careful use of pesticides, expanded irrigation, better farming practices and increased use of fertilizers all contributed to this result. FAO estimates that during the period 1985-2000 this pattern will continue in the developing nations.

2.2 Food supply

Population and agriculture

There are marked regional differences in the rate of population growth, with Africa's population increasing fastest, Figure 2.2.1.

Figure 2.2.1 EXPECTED REGIONAL POPULATION INCREASE BETWEEN 1985 AND 2000

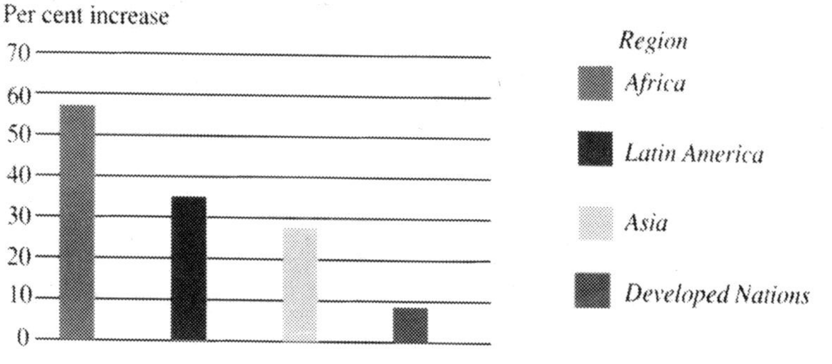

Source: UN (1986).

Food should preferably be locally produced. There are thus broadly corresponding regional differences in the need for increased food production.

The principal rural occupation is agriculture. It is the main source of livelihood for more than half of the world's population.

A high proportion of the world's most impoverished people are smallholders or landless workers in rural areas in developing nations.

Trade, industry and related activities are closely integrated and concentrated in urban areas offering employment and other possibilities that cause massive migrations to the cities, Figure 2.2.2. Consequent shortages of labour and skills occur in rural areas.

The large and growing city populations demand abundant and cheap food. Farmers need adequate and stable produce prices. This represents a continuous conflict of interest between the farming and the urban communities.

Figure 2.2.2 URBANIZATION BETWEEN 1950 AND 2000

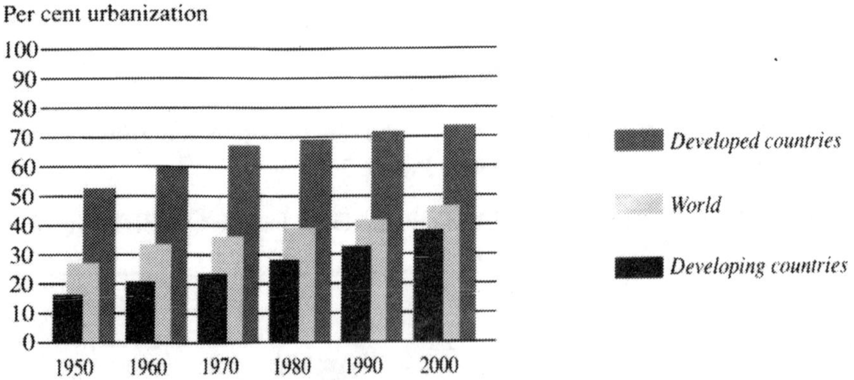

Source: UN (1986).

Subsidies and other economic and political decisions designed to encourage increased production through irrigation and the use of new varieties, fertilizers and pesticides also have social effects:

- mechanization has displaced labour
- increased need for capital borrowing restricts poor farmers and increases their vulnerability to fluctuations in prices and interest rates
- bumper crops depress produce prices and decrease rural incomes
- extensive pesticide use with insufficient guidance and control puts farmers, workers and the environment at risk.

The distribution of land ownership and the availability of capital and credit are often very uneven, and large land-owners are frequently the main beneficiaries of improved techniques. Such aspects of agriculture in developing nations have come in for severe political criticism. This criticism is also directed at agricultural techniques, including the use of fertilizers.

However, food production in many countries would have been much lower compared with present levels if irrigation, pesticides, new crop varieties and fertilizers had not been made available. Even more distress would have been the result.

Food needs

In the period between 1961/3 and 1983/5 world average food intake rose from 2320 to 2660 kcal/day per head, Table 2.2.1.

Table 2.2.1 FOOD AVAILABILITY

Food for direct human consumption, kilo calories per capita per day

Region	1961-63	1983-85
World	2320	2660
Developing countries	1960	2420
▪ Low income countries	1870	2310
▪ Middle income countries	2160	2660
Developed countries	3090	3370
▪ North America	3250	3630
▪ Western Europe	3110	3380
▪ Other developed market economies	2590	2890
▪ Eastern Europe	3160	3410

Source: Alexandratos (1988).

Earlier fears of extensive chronic food shortages proved unfounded. Human dietary needs range from 2000 - 4000 kcal/day depending on weight, age, sex, reproductive status (pregnancy, lactation) and physical activity. The food must contain protein; 70 g/day (11 g nitrogen/day) is generally adequate. If the protein is of high nutritional quality and easily digestible, the need may be lower, about 50 g/day. Earlier emphasis on protein need has diminished with the realization that protein needs will usually be satisfied when the dietary energy requirement is fulfilled (WHO 1985a).

A varied diet will usually satisfy the need for minerals and essential components such as vitamins. However, many eat monotonous diets, and vitamin deficiencies, e.g., of vitamin A, are not unusual in some developing countries. Local trace mineral deficiencies in the soil can give dietary deficiencies, e.g., goitre due to lack of iodine.

Statistics for average regional food availability conceal major differences among countries. Within a country dietary differences among the rich, the average citizens and the poorest can also be large. In some low income countries, food supplies per person in 1983/85 were no higher than those of 15 years earlier. There are estimates that some 350 to 500 million people remain seriously undernourished. The problem of chronic hunger has been solved for most of the world's population, but not for a significant minority.

Food production in different regions

There are marked differences among regions in the development of food production per capita in the last 20 years, Table 2.2.2.

Table 2.2.2 CHANGES IN FOOD PRODUCTION PER CAPITA

Basis: 1961-64 = 100

Region	Food production per capita, 1981-84
World	112
North America	121
Latin America	108
Western Europe	131
Eastern Europe & USSR	128
Africa	88
Middle East	107
Far East	116
Centrally planned economies in Asia	135

Source: World Commission (1987).

Parts of Africa have suffered a reduction in food production per capita due to unfavorable weather conditions, poor economies and war.

Increased incomes in industrialized nations have increased their demand for and production of meat and milk and have changed trade patterns for food and feeds. Meat production for export has increased in Latin America and Africa.

A major part of world grain production is used for animal feed. In principle the industrialized nations could feed substantially increased populations by adopting a more vegetarian diet.

The introduction of new technology has been uneven, and regional differences in the use of agricultural technology have increased. Thus the average food production increase (in t/ha) has been lower in parts of Asia and Africa than in Europe, Figure 2.2.3, but the relative yield increase (in per cent) in the last 20 years has been substantial in many developing nations. The "green revolution" in the tropics has been successful in raising food production though mainly in irrigated areas.

General reference: FAO (1987b).

Figure 2.2.3 GRAIN YIELDS IN SELECTED COUNTRIES 1965 - 84

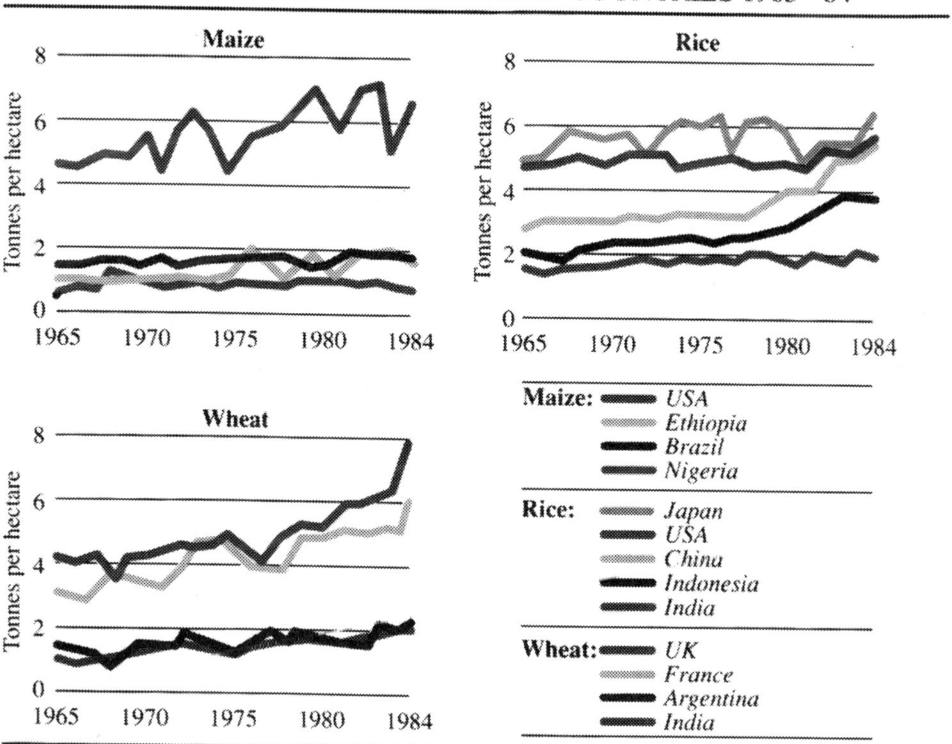

Source: World Bank (1986).

2.3 Limiting factors and challenges in agriculture

Factors that limit crop yields

Food production is a function of crop yield and cultivated area.

Crop yields are limited by a number of factors. Some of these, such as climate and length of growing season are largely beyond human control. Practices such as timely cultivation and seeding, selection of crop and variety, seed density and harvesting efficiency all influence final yield, but are not discussed further here. Other controllable yield factors are:

- land management
- water supply
- nutrient supply
- crop protection
- crop improvement through plant breeding.

Their management for increased crop production also gives rise to environmental issues.

The environmental impact and sustainability of farming

Arable agriculture is one of man's major changes of natural conditions.
A farm is an ecosystem very different from its natural state.

The issue, therefore, is not whether farming has an environmental impact, all farming has, but whether the extent of arable farming practices now causes unacceptable environmental change. The issue thus involves a judgement of benefits and costs. Though science can provide a factual basis, decisions about the acceptability of environmental impacts and agronomic practices are a political matter, involving factors others than those of science and economics alone.

A related challenge is the need for a sustainable agriculture: systems and practices that "meet the needs of the present without compromising the ability of future generations to meet their needs" (World Commission 1987).

Sustainable agriculture centres on protection, care and maintenance of agricultural land, with emphasis on prevention of soil erosion and degradation, careful husbandry of resources and environmental consideration.

A main theme of this book is sustainability and environmental protection in agriculture, and how improvements can be made through good practices.

Reference: FAO (1989).

Land

Man has caused major changes on our planet's land, notably in the last 100-150 years, Figure 2.3.1.

Figure 2.3.1 TRANSFORMATION OF LAND IN THE PERIOD 900 - 1977

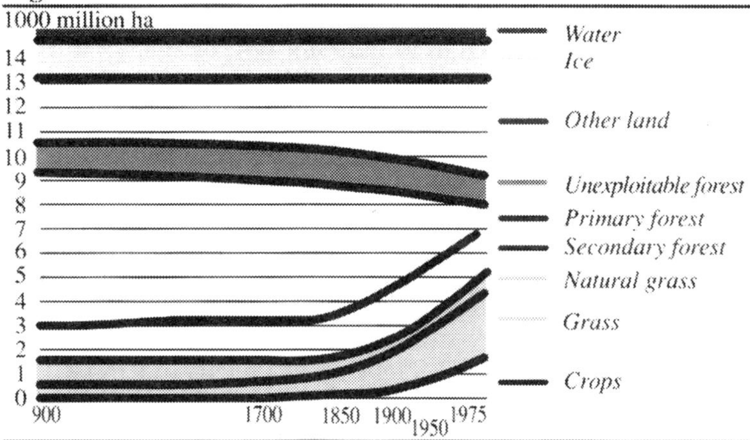

Source: Wolman and Fournier, SCOPE 32 (1987).

World land area is 14.9 thousand million ha, or 29 per cent of the total surface. At present about 10 per cent of all land is cultivated. A further 14 per cent is potentially cultivable.

The cultivated area is about equally divided between industrialized and developing nations. Developing nations harvest some 600 million ha annually. This is expected to increase by about 115 million ha by the year 2000 (83 million ha new land, and 32 million ha put into more regular arable use).

Cultivated area per capita varies greatly among regions (Table 2.3.1). The increasing population in some regions is putting agricultural productivity under severe strain.

Table 2.3.1 CULTIVATED AREA PER CAPITA

Region	Cultivated area per capita, ha 1964	1984
World	0.44	0.33
North America	1.05	0.90
Latin America	0.49	0.45
Western Europe	0.31	0.25
Eastern Europe & USSR	0.84	0.71
Africa	0.74	0.35
Middle East	0.53	0.35
Far East	0.30	0.20
CPE's in Asia	0.17	0.10

Source: World Commission (1987).

The developing nations as a group have large reserves of potentially cultivable land, but these reserves are unevenly distributed, Figure 2.3.2.

Figure 2.3.2 LAND USE AND RESOURCES IN DEVELOPING COUNTRIES

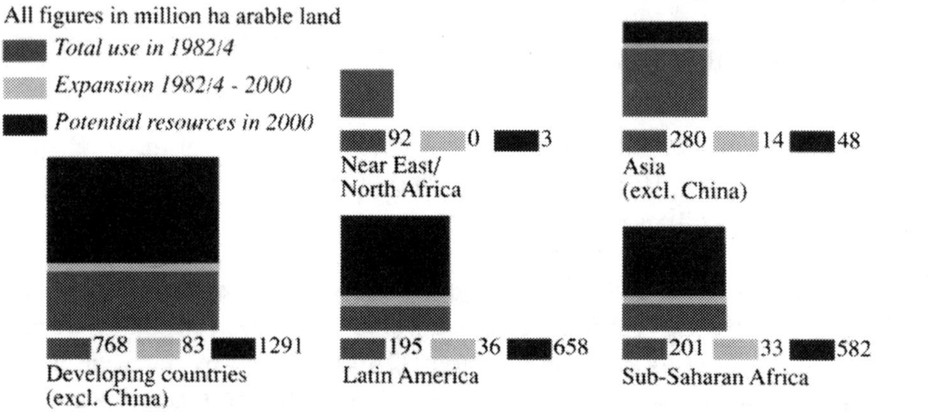

Source: Alexandratos (1988).

The figure shows:

1. *Arable land currently in use (brown)*
2. *Expected expansion toward 2000 (yellow)*
3. *Potential resources (blue).*

The squares in the figure are in relative proportion to the size of the respective land areas. The square to the left is the sum of the other four which represent different regions.

The reserves are mainly in Africa with around 580 million ha (of which Zaire has 160 million ha) and in Latin America with about 680 million ha (Brazil 520 million ha).

There are serious constraints on the use of much of these reserve areas for cultivation:

- poor agricultural quality of the land
- unreliable rainfall patterns
- sparse population
- conflict with environmental interests
- legal, social and economic factors.

In South and Southeast Asia, the Middle East and North Africa, land reserves for cultivation are scarce. North America and Australia should be able to increase their arable areas substantially. They are currently food exporters.

Western Europe has virtually no possibility of increasing its cultivated land area. Production in some regions can be increased by more irrigation together with greater use of fertilizers. Eastern Europe and the USSR should also have the potential for increased food production.

Destruction of forests

Destruction of tropical rain and highland forests is a major political and environmental issue. Forests contain valuable timber and the world's greatest numbers and variety of animal and plant species. They influence regional climate and water flow in their watersheds, and their destruction releases carbon dioxide (CO_2) into the atmosphere. Of the original 1600 million ha of tropical rain forests, only about 900 million ha remain. Much of the loss has occurred in the last 200 years, and the major part of this since the Second World War. Currently between 7.6 and 10 million ha are destroyed each year, and an additional 10 million ha are grossly disrupted.

Tropical rain forest in Costa Rica.
Photo: O. B. Frøshaug, Samfoto, Oslo.

The tropical rain forests are distributed between:

- Latin America, 57 per cent
- Southeast Asia, 25 per cent
- Africa, 18 per cent
- others, 10 per cent.

In all these regions deforestation takes place, though statistical data are uncertain.

Forests on prime land are especially valuable ecosystems but are also at highest risk. Forest clearing for the period 1975 - 2000 is expected to affect:

- about 15 per cent of total forest area
- about 55 per cent of forests on land with potential productive agricultural land
- about 70 per cent of forests on highly productive land.

Forest decline has several causes:

- logging without replanting
- mining without land reclamation
- establishment of large cattle ranches
- landless farmers seeking new land
- increased traditional shifting agriculture, where the fields are cleared by burning, cultivated for a few years until topsoil nutrients are exhausted, then left for years for the forest to recover and reestablish the topsoil nutrient reserves. Increasing population pressure can shorten this cycle beyond the recuperative power of the system.

Other forests are also under threat. Firewood remains a major energy source for cooking and heating in the developing nations. The increase in population and intensive use of land for cultivation causes firewood to be in increasingly short supply, and forest degradation results.

The developing nations face major internal social and food supply problems and a pressing need for conservation of the remaining undisturbed forests, wetlands and plains. Larger yields on land already in cultivation could help ease the pressure for land development. Adequate fertilizer supplies could increase indigenous food production and be a part of the solution.

Reference: FAO (1986).

Problem soils

In the tropics and subtropics, some areas have fertile soils but the soils are often very different from those encountered in temperate regions such as Northwestern Europe. Tropical and subtropical soils are often geologically old and have been exposed to leaching for very long periods of time. They can therefore be deficient in both major and minor essential elements. The high temperatures induce rapid decomposition of soil organic matter, and the soil can have poor physical structure.

Some soils are very acid or alkaline or have a high salt content, all factors that are detrimental to plant growth. Others consist of coarse sand with low nutrient content; some contain minerals that easily form sulphuric acid when the soil is cultivated.

Special skills and costly inputs including fertilizers are often required when these soils are cultivated.

Successful management of tropical and subtropical acid soils is a major challenge. The problem is not only one of soil quality, plant breeding, nutrient supply, crop protection and agronomic techniques, but also one of land tenure systems, finance, logistics, marketing and education.

Reference:
Webster and Wilson (1980).

Land losses

Loss of cropland is a cause for serious concern, but the extent of loss can at present only be estimated. There are large regional differences, but few countries have reliable data on the subject. The problem is not only one of quantity, but also of quality: often the best soils are at the highest risk. Current estimates (Wolman and Fournier 1987) are that global land losses for the years 1975 - 2000 will be:

- about 4 per cent of all potential productive agricultural land
- about 25 per cent of all highly productive land.

Losses are largely due, directly or indirectly, to mis-management.

The shift in priorities permitting the use of productive land for roads and urban facilities is an example. Estimates are that 400 million ha will be used for non-agricultural purposes by the year 2000. Much of this was formerly prime agricultural land.

Inappropriate agricultural practices can accelerate land degradation and loss, e.g.:

- arable cropping on easily erodible land
- tilling at the wrong time or leaving the land fallow
- irrigation without preventing salt accumulation or waterlogging
- overgrazing
- soil compaction through use of heavy machines.

Correct practices can conserve and improve soil quality. This subject is discussed in more detail in Chapter 4.

References:
Alexandratos (1988), Wolman and Fournier (1987).

Water

Water is the principal constraint on crop yield on much arable land. While rainfall is adequate for crop needs in some regions, others have erratic or sparse rain. Irrigation practices have a long history.

Lowland rice requires flooded fields. Rice planting in India.
Photo: Jarle Rognaldsen, Oslo.

Currently 15 per cent of the world's total arable land is irrigated, but this produces 36 per cent of the total crop yield. Irrigation both increases and ensures yields and may permit harvesting of 2 or 3 crops per year instead of 1 or 2.

Crops must have adequate water supply in order to utilize nutrients properly. Where growth is severely water-restricted, fertilization is of limited value. Water and nutrient management is, therefore, connected, and regions where irrigation is common also tend to use more fertilizers than is usual in areas only watered by rain.

Part of the required increase in food production may come from an extended use of irrigation with associated nutrient use provided that water of sufficient quality is available. This could provide for about two thirds of the increase in arable land up to the year 2000, and the proportion of irrigated land could increase to about 20 per cent of total arable land. Irrigated agriculture is most common in Asia where 85 per cent of the expansion of irrigation is expected to take place, most of it in India. The greatest potential for expansion in irrigation after the year 2000 is in Latin America, Figure 2.3.3.

Figure 2.3.3 IRRIGATED LAND IN DEVELOPING COUNTRIES

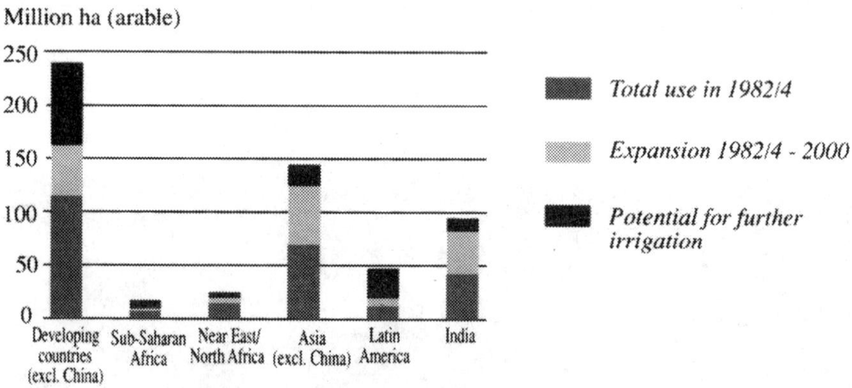

Source: Alexandratos (1988).

The figure shows:

1. *Arable land currently irrigated (red)*
2. *Projected expansion toward 2000 (yellow)*
3. *Potential for further irrigation (blue).*

Irrigation in Li Jiang Valley, Peoples Republic of China.
Photo: Arne Walderhaug, Samfoto, Oslo.

Only a fraction of the areas where water is growth limiting at present can be irrigated in the future, either due to high costs or to lack of water. Global water use increased 350 per cent between 1950 and 1980, and the increase continues. This increase is due to expanding populations and growing water use in agriculture and industry. Agriculture accounts for about 73 per cent of all water use. Regional water shortages are increasing.

Irrigation can be expensive, its management complex and equitable water distribution difficult to achieve, particularly in large scale irrigation projects. High skills are required to prevent soil salinity, maintain irrigation networks and pumping stations, prevent water loss, and to distribute and use the water in an efficient manner.

Irrigation benefits crop production but water management poses severe challenges:

- excessive use of ground water is in some regions (e.g., USA) depleting these resources
- filling up of water reservoirs with sand and silt severely threatens the viability of important irrigation and hydroelectric power facilities

- salts in irrigation water can accumulate in the soil. The salt can be removed by drainage, but disposal of brine is a problem
- pollution reduces water quality
- waterlogging reduces crop yields and can be detrimental to soil quality. Drainage is a traditional counter measure
- soil erosion by water is a serious issue, see Section 4.7
- floods cause extensive damage
- deforestation of rain catchment areas in the mountains can increase the frequency and destructiveness of flooding. It is now common to protect such areas by legislation.

Improved and extended water control and use remains one of the main challenges in agriculture.

Reference: James (1988).

Plant nutrient supply

Crop growth and yield are governed by environmental conditions and the availability of water and nutrients. Since the Second World War, world average cereal yields have increased from about 1.2 t/ha to 2.5 t/ha at present (Stapel (1982)). This increase arises from use of fertilizers, plant protection, irrigation and new plant varieties, Figure 2.3.4.

Figure 2.3.4 GLOBAL TRENDS IN POPULATION GROWTH, GRAIN YIELD, AND ORIGIN OF PLANT NUTRIENTS

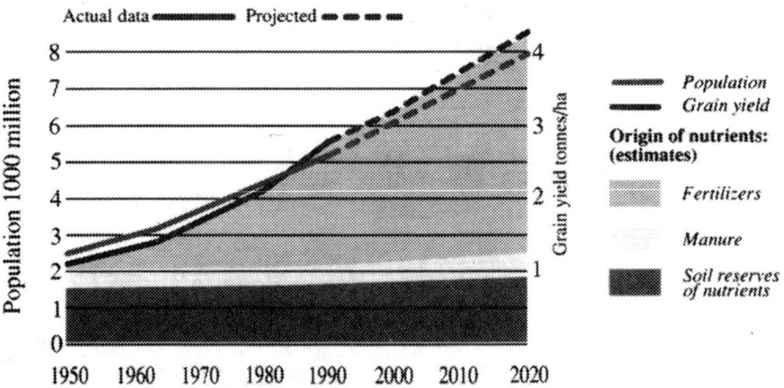

Source: Stapel (1982) with latest data added.

The soil provides the principal natural source of nutrients. Return to the soil of nutrients in manure and organic wastes, and input through nitrogen fixing crops (legumes) contribute additional nutrients to crops, but supplies from these sources have increased only marginally. Fertilizers provide the additional input required for increased crop yields (Cooke (1982), Olsen (1987)). This yield increase has closely followed the increase in world population and must continue to do so in the future if hunger is to be avoided.

Today fertilizers support about half of the world's grain production. Other factors are also important, but adequate nutrient supply is essential for high yields. Long term experiments with and without fertilizers in many countries confirm this. Those at Rothamsted (UK) are an example, Figure 2.3.5.

Figure 2.3.5 GRAIN YIELDS WITH AND WITHOUT FERTILIZERS

The Rothamsted experiments

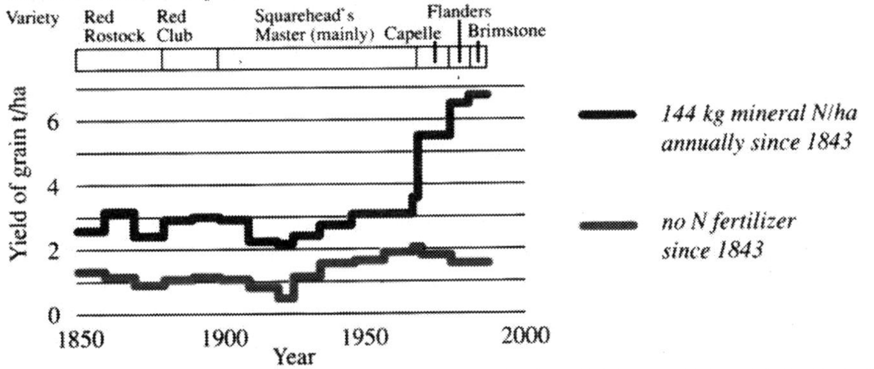

Source: Jenkinson (1982, 1988/89).

These experiments illustrate the long-term viability of fertilizer use in maintaining yields, the dramatic yield increase since 1969 due to modern crop varieties which are responsive to fertilizers, and the protection of yield potential with herbicides and fungicides. On unfertilized land these varieties give a markedly lower response in absolute terms.

The results reflect practical experience with increased yields in Western Europe in this century, Table 2.3.2.

Table 2.3.2 DEVELOPMENT OF SOME ANNUAL NATIONAL AVERAGE CROP YIELDS IN EUROPE

	Winter wheat UK	Sugar beet France	Maize France	Hay (dry matter) FRG
Year	t/ha	t/ha	t/ha	t/ha
1900	2.0	26.0	1.1	3.9
1930	2.0	33.0	1.7	4.6
1950	2.6	34.0	2.2	4.7
1970	4.2	45.0	5.1	6.8
1988	6.2	60.0	7.2	8.1

High yields can also be achieved without fertilizers if manure and crop waste are used in sufficient amounts. In some areas, for example the Netherlands, manures could largely satisfy the total agricultural requirement for nutrients. However, this situation arises only because large amounts of feed, grown elsewhere, are imported to support intensive animal production. Fertilizers are needed to produce this imported feed. Our estimate is that today in Northwestern Europe as a whole, improved use of manure could replace about 10 per cent of the fertilizer currently used.

Fertilizer use and supply

World fertilizer use in 1987 and the estimated needs in 1995 are:

Table 2.3.3 WORLD FERTILIZER DEMAND

(million tonnes per annum):

Nutrient	1987	1995
Nitrogen (N)	72	91
Phosphorus (P)	15	18
Potassium (K)	22	26

Source: Bumb (1989).

Increased cultivation of legumes can supply only a minor part of the required increase in nitrogen input. Legumes, however, require space at the expense of other crops, and are less efficient as nitrogen sources than fertilizers. The use of legumes as a nitrogen source is discussed in Section 5.3.

The rate of fertilizer use varies greatly among regions, Table 2.3.4, and the various crops differ in their share. In some regions, e.g., in Africa, cash crops (e.g., cocoa, coffee) are better fertilized than are food crops.

Table 2.3.4 REGIONAL FERTILIZER USE

	Kilo nutrients applied per hectare kg ($N + P_2O_5 + K_2O$)	
Region	1964	1984
World	29.3	85.3
North America	47.3	93.2
Western Europe	124.4	224.3
Eastern Europe & USSR	30.4	122.1
Africa	1.8	9.7
Middle East	6.9	53.6
Far East	6.4	45.8
Latin America	11.6	32.4
CPE's in Asia	15.8	170.3

Source: World Commission (1987).

Most of the increase in global fertilizer production in the 1980s was in Asia and the USSR.

Developing nations face major challenges in producing and distributing the fertilizers required for increased food needs. Local sources are preferable, but these are often not available in adequate amounts or quality. Fertilizer or raw material imports are therefore required.

Fertilizer aid can supplement local production. This is further discussed in Section 2.4.

Fertilizer plants are complex to build and operate and involve heavy investments. Hence the need for large production units, and for skilled operators. Lack of transport facilities can cause great problems in fertilizer distribution, resulting in farmgate shortages.

Besides difficulties with financing fertilizer production and purchase, insufficient knowledge about proper fertilizer use for local conditions and insufficient training of advisory personnel and farmers also limit efficient fertilizer use.

References: Bumb (1989), FAO (1987a), IFA (1986).

Crop losses through pests: weeds, diseases and insects

Crop damage through pests is a major constraint on food production:

- weeds compete with crop plants for sunlight, water and nutrients
- insects and their larvae feed on crop plants and act as disease carriers
- fungi, bacteria and viruses invade and damage the plants
- rodents and insects feed on produce, and fungi poison them or make them otherwise unsuitable as food.

Occasionally the attacks become national or continental disasters: beginning in 1845, blight (a fungal disease) destroyed the Irish potato crop; a root louse ravaged European vinyards between 1870 and 1900; blight destroyed much of the US corn crop in 1970; and lethal yellowing (an insect borne disease) is currently devasting coconut plantations in the Caribbean. Such disasters are not limited to cultivated fields; Dutch elm disease and the African locust problem are recent examples.

Disasters of this magnitude are not common, but severe pest problems are usual on regional or local levels. Yield and quality reductions occur even when the attacks are slight or well controlled. Measures to protect crops have always been an important part of farming.

A wide variety of chemicals is now used to help control weeds, crop diseases and other pests. Agrochemicals (originally copper and sulphur preparations) have been used for more than a hundred years, but their present development dates from discoveries made at the time of the Second World War. Many current agrochemicals are complex organic compounds applied

Left: Severe mildew attack on barley. Large reduction in yield can occur if crop is not protected with fungicides. Photo: Ciba-Geigy.

Right: Insects can damage and destroy crops. A heavy infestation of aphids in wheat. Photo: ICI.

in water solution. They are an integral and indispensable tool in current agriculture though there remain important plant diseases (e.g., take-all in wheat) for which no protective pesticide is available.

New, improved products are being developed, but the cost of research and testing to meet requirements for registration for a new compound is estimated to be more than USD 50 million and takes six to ten years. The cost is even higher when unsuccessful projects are taken into account. The introduction of new pesticides has therefore diminished to about 2-3 per year at present. The present trend is to reduce agrochemical use through increased attention to field conditions.

The use of agrochemicals is controversial. There are many issues, e.g.,

- the toxicity and environmental persistence of some of the products
- risks in production and distribution
- risks to farmers and workers
- development of pest resistance
- concern about residues in produce.

These are important issues, but outside the scope of this book. The topic was recently reviewed by Danish authorities (Danish Government Plant Protection Centre (1986)). Crops produced by current intensive agriculture can have increased vulnerability to pests. This is further discussed in Section 4.9.

Efforts to replace chemical pesticides by mechanical and biological control methods have had somewhat mixed success, but the search for such alternatives is continuing. The practical difficulties have been many and progress has been slow. Any new method for pest control will start natural selection in the pest population. This in time results in development of adaptation or resistance by the pest. The quest for pest control methods is therefore an unending challenge. There are no easy solutions.
Reference: Agrios (1988).

Weeds in a barley field in Norway. If weeds are not controlled, crop yield is greatly reduced and harvesting made difficult.
Photo: H. Hansen, Norsk Plantevernkjemi.

Plants: their growth and stress resistance

The capacity of plants for growth and development is the ultimate yield constraint.

Plant growth rate is not uniform: it starts slowly, then accelerates as leaf area increases. When the field canopy closes, the crop cannot further increase its sunlight interception, and then growth rate remains stable until maturity and senescence. During this period the crop passes through a series of development stages. This is illustrated for cereals in Figure 2.3.6.

Figure 2.3.6 CEREAL GROWTH STAGES AND NUTRIENT UPTAKE

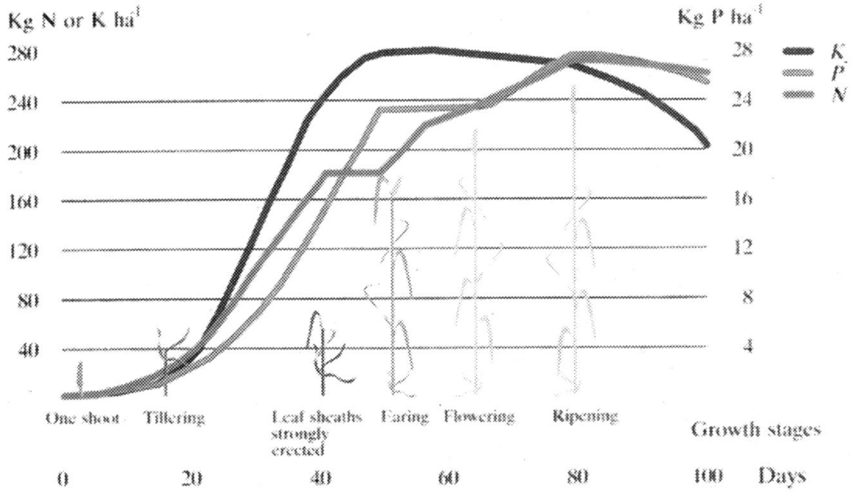

Source: Nielsen (1983).

Nutrient uptake by crop plants varies throughout the period of growth. Rapidly growing cereals take up nitrogen and potassium at a rate of about 5 kg/ha • day and of about one tenth of this rate for phosphorus (Barraclough (1989)). Most of the nutrients are taken up in the early part of the growing season. The length of the period of most active nutrient uptake varies with the crop and also to some extent with variety. During maturation, nutrients and carbon are moved from other parts of the plant to the grain, seed or tuber (Spiertz and de Vos (1983)).

About 20 - 30 tonnes of dry matter per ha above ground can be produced in a season under European conditions. The amount depends on crop species and length of growing season.

A wheat field yielding 10 tonnes grain/ha actually produces some 17 tonnes dry matter/ha. There may be some potential for further yield increase, but it does not seem large.

Plants differ in their efficiency in utilizing sunlight energy, carbon dioxide and water. Some plants originating in tropical areas (e.g., maize and sugar cane), are usually more efficient and fast-growing than the common temperate crop plants such as wheat and barley by some 30 - 50 per cent. This is due to differences in plant anatomy and biochemistry.

Growth and development will be reduced at any development stage if growth factors such as nutrients or water are not available in required amounts. Deficiency of nutrients such as nitrogen gives smaller leaves and reduces the capability for sunlight interception. Fewer ear-bearing side-shoots (tillers) will develop in cereals.

Grain yields are determined by the factors:

number of plants/ha • ears/plants • grains/ear • grain weight

Plant density is determined by seeding rate and survival. Too large spacing between plants prevents or delays canopy closing, too dense planting causes mutual hindrance and reduces yields. Fertilizers influence and enhance the three other yield factors, but the relative effect on these varies with the time of application, field conditions and the variety.

Plant breeding has contributed greatly to the high yields of present crops. Before the World Wars, wheat varieties in UK stood about 100-140 cm tall; today, they stand at about 75 - 90 cm. It is not always the case that the smallest cultivar among the modern varieties gives the greatest yield. But in general the present varieties produce less straw and correspondingly more grain, and more of the growing season is used for grain filling. Straw stiffness has improved with reduced risk of lodging, and modern crops can carry increased ear weight and utilize more nutrients. Lodging can reduce yield and quality and make harvesting difficult. Agrochemicals which enhance straw stiffness are available, but their use is not universally permitted.

The potential for even more efficient varieties in the future is not exhausted. However, improved adaptation to local conditions, increased vigour and resistance to stress factors (e.g., fungal diseases) seem to offer greater potential for future improvements than further basic changes in plant size.

Gene transfers and other new biological techniques: new tools for plant breeding

Systematic plant breeding was given a scientific basis early in this century through the discovery of laws of heredity and of genes as trait carriers. Special techniques were developed which made it possible to surmount barriers between species and permitted crosses between crop plants and distant relatives. For example goatgrass was crossed into wheat as a source of resistance to eyespot (a fungal disease). As new discoveries and techniques became known, plant genetics and breeding evolved as a complex and fascinating science and art. This rapid development continues, and will remain the mainstay of plant breeding for the foreseeable future.

In addition, discoveries in molecular genetics are now furnishing new tools with great potential for future plant breeding. These techniques are based on the elucidation in 1953 of the molecular basis of inheritance. Methods were developed for detailed studies of genes and how they work, and for their transfer to new hosts.

The first genetic transformation of a plant was done as recently as 1983. Techniques have developed rapidly, and transformations and genetic studies are now done routinely in academic and major plant breeding laboratories, both in the industrialized and the developing world.

There are, however, technical constraints on the present use of these techniques. Cereals are difficult though not impossible to transform. Only single or a few genes can be introduced at the same time, while many traits depend on the coordinated expression of numerous genes. It is difficult to find and isolate interesting genes among the multitude present in nature, and the techniques and their applications demand a high level of skill and insight. Even so, the new methods open possibilities for breeding plants with improved qualities and resistance to various adverse environmental factors: insects, diseases, salt and drought.

Most of these are classical breeding objectives, but modern genetic techniques should improve the precision in breeding and accelerate the process, greatly extending the range of genes available to the breeder.

It may even be possible to develop crops with biological nitrogen fixation, or with improved efficiency in nutrient uptake and use, such as cereals with reduced amounts of nitrogen in their root mass at harvest.

However, plant breeding remains a slow process: it takes at least 5, more typically 10 years or more, from the first crossing to the day the variety can be presented on the market. Thus, while progress in the laboratory is rapid, today's results will not appear in strength in the fields until well into the next century.

Rapid advances in biochemistry are giving detailed knowledge of the causes of crop damage due to pests and diseases. This is likely to lead to more efficient crop protection measures.

Fears are being expressed that the new techniques are dangerous, that they may lead to environmental disasters and give unfair advantages to advanced countries and companies, and that the techniques and their application must be strictly controlled or even banned. But the prospect of using genetic techniques for solving pressing agronomic problems in the developing nations seems too promising and the needs too great to forego the controlled use of these new tools in plant breeding.

Performing crosses in the traditional manner and developing new varieties is very time consuming. As long as the plant breeder is limited to conventional techniques alone, only a fraction of the available genetic material can actually be used in practical breeding.

The new genetic techniques greatly extend the range of genes that can be used. Thus preservation of the large pool of genes present in nature is now a matter of increasing importance. The need for preservation of ecosystems in their natural state, especially forests and grasslands, should therefore be stressed.

References: Grierson and Covey (1984), Tudge (1988).

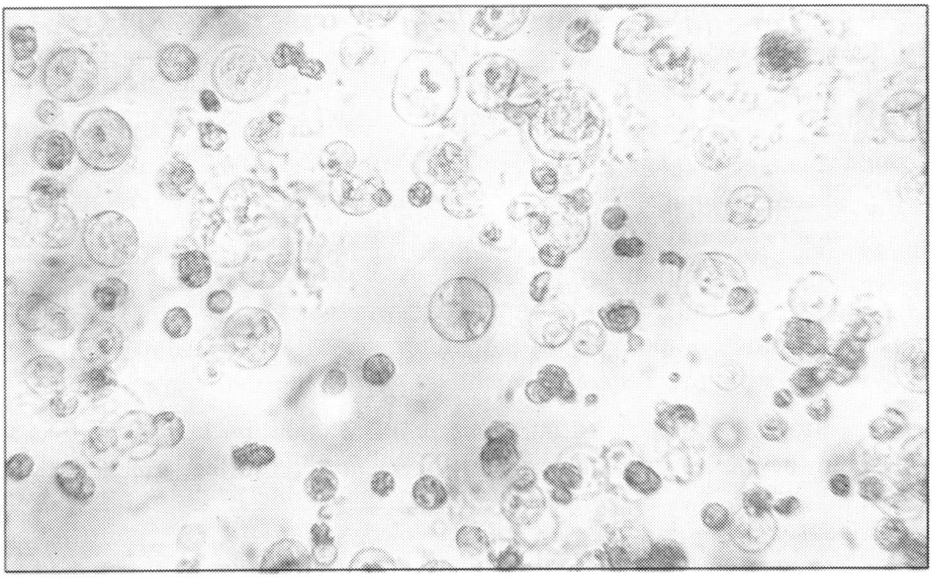

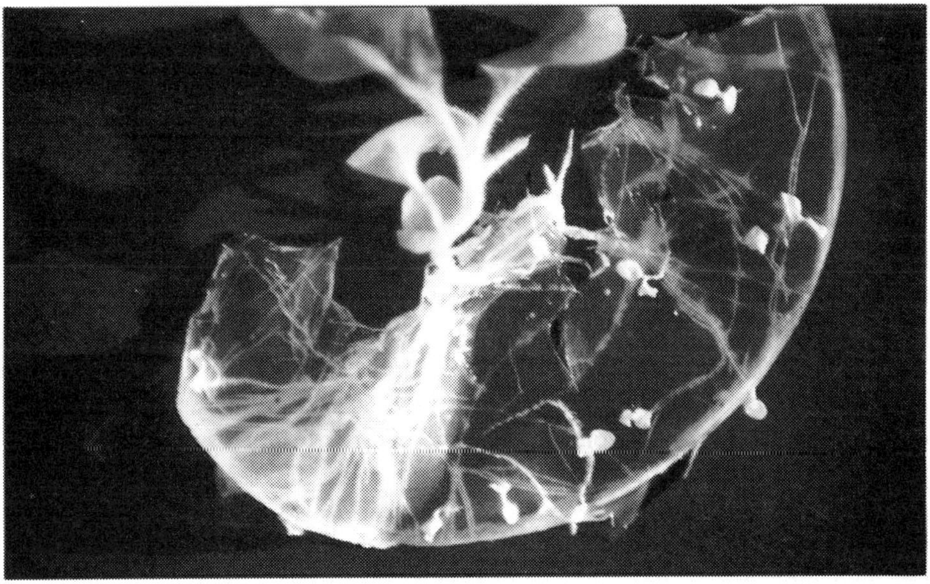

Various methods are available for gene transfer into plants. In one procedure, cells are taken from plant tissue and grown as cell culture. Chemical and physical treatment make the cell wall permeable to genetic material. Specially constructed genes from other organisms are added and a few cells develop into plants under conditions where the new genes are essential for growth, and where non-transformed plants wither.

Above: Plant cells in culture. Photo: R. Potter, Agricultural University of Norway.
Below: Only transformed plants grow in the selection procedure. Photo: R. Aalen, Plant Molecular Biology Laboratory, NLVF, Ås, Norway.

Weather and climate: the greenhouse effect, the ozone layer, and agriculture

Agricultural yields are very dependent on weather. Recent examples are frost and drought damage to Soviet grain crops, the recent drought in the USA and frost damage to Brazilian coffee. There are indications that extreme weather conditions now occur more frequently than in the past. If true, this instability will influence agriculture and increase the need for food reserves. Unfortunately, accurate long term weather forecasts are probably impossible. Some compensation for weather vagaries is possible by irrigation, but apart from this little else can be done with the weather.

One aspect of the climate now attracting political attention is the probable increase in the earth's average surface temperature due to the "greenhouse effect".

The heat balance of the atmosphere is influenced by some of the components of the air that are present in only small amounts: carbon dioxide (CO_2), methane (CH_4), nitrous oxide (N_2O) and the chlorofluorocarbons (CFC's). These gases absorb infra-red (heat) radiation from the earth. The concentration of these gases in the air is increasing and thus the capacity of the atmosphere to retain heat. It is feared that this will give a global temperature rise and thus influence climate.

Ozone is formed in the upper part of the atmosphere through the action of intense sunlight. The trace gases nitrous oxide, the chlorofluorocarbons and some related compounds are chemically stable in the lower part of the atmosphere, but react in the ozone layer. These reactions consume ozone. It is therefore feared that increasing concentrations of these gases will reduce the effectiveness of the ozone layer as a protective shield for the earth against ultra-violet radiation (UK Stratospheric Ozone Review Group (1987)).

The increasing atmospheric concentration of these gases is due to human, domestic and industrial activities. The increase appears to have started with industrialization and became especially noticeable after the Second World War.

Table 2.3.5 GREENHOUSE GASES

Gas	Surface air concentration ppm (vol) Year 1600	1880	1985	Recent growth per cent /year	Expected relative contributions to global warming (1980-2050) Per cent
CO_2	280	280	345	0.2-0.5	47
CH_4	0.7	1.15	1.65	1.2	14
N_2O	0.28	0.28	0.3	0.2-0.3	10
CFC's	0	0	0.0005	5.8	29

Based on: Bouwman (1990) and Gushee (1989).

The influence of these gases on global temperature and climate is very complex. Other factors are also involved, such as regional pollutants (e.g., ozone), water vapour and clouds, and changes in the earth surface reflectance due to deforestation and other land use changes.

There is a growing consensus that the increase in the concentration of greenhouse gases gives cause for concern. The observed trends must eventually change the global surface temperature or the atmospheric circulation pattern, or both, but the size and timing of any effects are a matter for current debate.

In principle the emissions of chlorofluorocarbons should be those easiest to reduce or eliminate. These are man-made industrial chemicals. There is now international agreement that their use should be greatly reduced and preferably eliminated.

The main concern is the increasing concentration of carbon dioxide (CO_2.) Burning of fossil fuel provides an annual carbon dioxide input to air of about 5000 million tonnes of carbon*. This amounts to 0.7 per cent per annum of the atmospheric content of 715 000 million tonnes of carbon. Further details about present use of fossil fuel are given in Section 4.11.

* *To ease comparisons between carbon dioxide in the air and carbon in fuel, forests and soil, carbon dioxide is given as carbon equivalent.*

The world's forests contain some 480 000 million tonnes of carbon. When trees are replaced by grassland or crops with less standing biomass, carbon is liberated as carbon dioxide. The present rate of decline in forest area results in a substantial input of carbon dioxide to the atmosphere, though somewhat less in magnitude than the burning of fossil fuels.

Soils also contain much carbon, about 800 000 million tonnes, as humus. Increased conversion of grasslands to croplands liberates some carbon from this source too.

The increase in atmospheric concentration of carbon dioxide has increased plant uptake by some 5 - 10 per cent. At present about 40 per cent of the carbon dioxide emitted to the atmosphere is taken up by the oceans. Neither of these factors can compensate for the large scale emissions, and the future atmospheric concentration is expected to rise to about 440 - 660 ppm by the year 2050 unless effective steps are taken to reduce such emissions.

The use of fossil fuel is on so large a scale that it cannot be supplanted by biomass burning. Apart from encouraging energy efficiency in agricultural practice (Section 4.11), the main contribution of agriculture and the fertilizer industry to abating the carbon dioxide problem must be to maintain soil humus and to increase the productivity of arable land and thus reduce the need for using forested areas for crop production.

Methane (CH_4) is a gas with a rather short atmospheric residence time of about 8 years. It is converted to carbon dioxide through various oxidation processes in the atmosphere.

Agricultural practices, mainly water management and animal husbandry, are believed to be major factors in the current noticeable increase in methane production.

Sources of methane are mostly anaerobic fermentation of organic matter by ruminant cattle, and also in wetlands such as rice fields. Increasing rice - paddy areas could be a major cause for the increased emissions. About 20 per cent of atmospheric methane seems to be derived from fossil sources.

Nitrous oxide (N_2O) comes mainly from the soil. Nitrous oxide is a greenhouse gas and also reacts with ozone in the ozone layer, where it is decomposed. The air contains some 1500 million tonnes of nitrogen in the form of nitrous oxide. The annual input was some 10 million tonnes, but this has increased in the last decades to about 15 million tonnes per year. The annual increase in concentration is only about 0.25 per cent, but the long residence time in the atmosphere (over 100 years) makes even the present increase a subject of concern for the long term.

Highly fertile cultivated land in temperate and tropical regions together with tropical forests and grasslands are regarded as the major sources of nitrous oxide, but the relative importance of the individual sources and processes is largely unknown. Combustion was believed to be a major source, but this contribution seems to have been overestimated.

Nitrous oxide is formed in the soil during the natural biological processes of nitrification and denitrification. The end-product of denitrification is usually nitrogen gas but under some conditions nitrous oxide is also formed.

The emission of nitrous oxide from fields is uneven, it depends on competing reactions and varies with the conditions.

Agricultural practices such as tillage, fertilization, manuring, crop residue management and drainage all influence nitrogen transformation processes in the soil and may therefore influence nitrous oxide emissions. Knowledge required to give guidelines on how to minimize such emissions is largely lacking. Nitrate application on excessively wet or waterlogged fields can increase nitrous oxide emissions from the soil, but it is not usual to fertilize under such conditions. Research on the influence of agriculture on nitrous oxide formation is now being undertaken in many institutions, but improved measurement methods for field emissions are needed.

References: Bouwman (1990) Warneck (1988).

Food losses after harvest

There are estimates that frequently some 10 - 30 per cent of harvests are lost through:

- damage or consumption by insects and rodents
- bacterial and fungal damage
- quality deterioration and losses in transport, processing, distribution and use.

Efficient storage techniques are available, but generally imply heavy investment, operation of advanced technology and reliable energy supply (e.g., cold storage under controlled atmospheric conditions). The wider application of known technology should have some potential for stretching food supplies.

Reference: Lieberman and Coursey (1983).

2.4 Food trade

Most food is produced and consumed locally or regionally. International trade in wheat and maize is some 15 - 20 per cent of world production, that of rice modest in comparison. Agricultural production has increased more in some regions than in others and the pattern of world cereal trade has shifted, Figure 2.4.1.

Figure 2.4.1 TRENDS IN FOOD TRADE AND TRADE BALANCE 1961 - 84

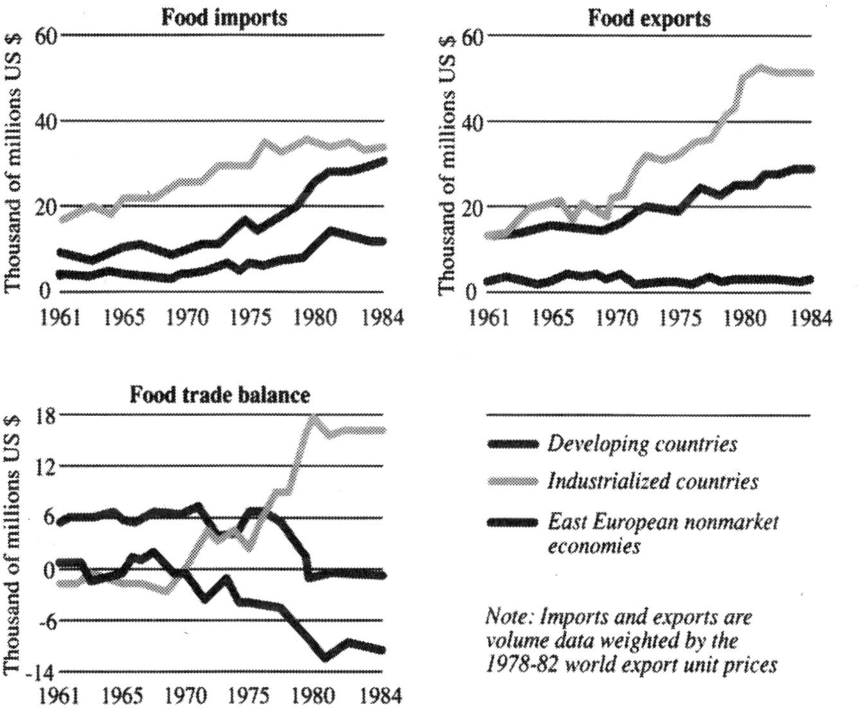

Source: World Bank (1986).

North America exported less than 5 million tonnes annually before the Second World War, but close to 120 million tonnes annually during the 1980s. The 1970s were a period of major change. Western Europe as a whole is now exporting, while it had considerable shortages immediately after the Second World War. China, Japan and the Soviet Union bought half of the world's grain exports early in the 1980s. The rest was bought mainly by relatively rich developing countries, such as oil exporting countries in the Middle East. Many African countries south of the Sahara became net importers of cereals, and are partly dependent on aid.

The dominant factors behind the change in trading patterns were growing populations in the developing countries, stagnant Eastern European production (at least partly due to adverse weather), and increasing agricultural productivity in the US and in the EC countries.

The general picture conceals large regional differences. The developing countries are a very variable group, with poor nations such as Bangladesh and upper-middle income countries such as Brazil, each with different problems and potentials. Some of the developing nations (e.g., Argentina) are major food exporters, while heavily industrialized Japan is a net importer. The USA was and remains the major food exporter; it still has a large and unused potential for further expansion of production and export.

Global stocks of food grew during the 1980s, Table 2.4.1, especially in North America, to the equivalent of 27 per cent of world annual consumption in 1987. This was followed by a dramatic fall. The estimated reserves in 1989 are now below the FAO guide-line for the stockpile required for world food security (18 per cent of annual consumption). This illustrates the effect of the 1988 drought, and of the sensitivity of world food supply to crop failure in the major producing countries.

Table 2.4.1 STOCKPILES OF MAIZE, WHEAT AND SOYBEANS

Millions of tonnes*

Year	Maize		Wheat		Soybeans	
	USA	World	USA	World	USA	World
1988/9	49	87	19	117	5.0	18.1
1987/8	116	145	34	144	8.3	19.8
1986/7	133	162	49	175	11.9	19.7
1985/6	110	144	52	168	14.9	23.2
1984/5	45	89	39	164	8.6	17.6

* USA figures are as of Sept. 1 for maize and soybeans, Jun. 1 for wheat. World-wide figures are as of Oct. 1.

Source: US Department of Agriculture.

Agriculture is integrated into the overall world economy. As a result, international economic conditions have a powerful influence on agriculture in most countries. Instability, debt, recession, fluctuations in exchange rates, good commodity prices followed by a slump: all inevitably affect agriculture. During recent years, heavily indebted countries with a high dependence on agricultural exports have been badly affected, especially in Latin America and Africa. However, local demand and production have been maintained except in the worst affected countries.

In the 1970s the growing import demand in middle income countries was met by increasing production and export by developed countries. After 1980 there was a sudden halt in the growth of demand, despite increasingly depressed prices. This caused stagnation in world agricultural trade. International prices reached their lowest levels since the Second World War in real terms. Prices also fell for traditional tropical export crops due to overproduction.

Food and fertilizer aid

At present direct food aid to poor countries with low production is an important part of their food supply. Such nations depend on world trade and stocks in order to overcome recurrent crises, e.g., droughts. Food aid has accounted for 14 - 24 per cent of the cereal import of low income, food deficient countries. It played a critical role in relieving the African famines, but aid keeps local prices down and can discourage local food production.

Aid is also given in the form of funds for fertilizer purchase on the international market, but the scale is small compared with the level of food aid. In 1983, the Sub-Saharan nations received USD 540 million in food aid, USD 26 million in fertilizer aid and USD 22 million for other agricultural inputs. Even with later increases (USD 127 million for fertilizers in 1985), the difference is substantial, although in many instances fertilizer aid would be more appropriate and effective than food aid.

Fertilizer aid can help in situations of fertilizer shortage where local production cannot satisfy the needs. Such situations are common in large regions (e.g., Africa).

Fertilizer aid can help to increase local food production and stimulate rural development. Fertilizer imports do not strain scarce local transport systems to the extent that food aid does: generally 1 tonne of fertilizers gives around 10 tonnes of grain (FAO (1987a)).

To provide its full potential benefit, fertilizer use should be supported by advisory services backed by adequate research and trials, and by the availability of healthy seeds and crop protection agents. Even with this support, fertilizers will only be effective if there are no overriding natural barriers to crop growth, e.g., lack of water.

Fertilizer aid is no panacea - like all aid it is not a satisfactory permanent solution to regional food shortages, social and economic problems. But in the current situation in many developing nations, it plays an important role in development.

General references to Chapter 2: Clark C.W. (1989), World Resources Institute (1987), Worldwatch Institute (1989).

Aid for fertilizer import can support local food production in developing nations. Photo: T. Bergli, Norsk Hydro.

CHAPTER 3

AGRICULTURE IN WESTERN EUROPE

Agriculture is not static, it is continuously evolving. The present debate on environmental consequences of agriculture and its balance of costs and benefits should also be seen in a historical perspective. The development of agriculture in Western Europe is an interesting case in point.

3.1 The development of Western European agriculture since the Second World War

Agricultural policies since the Second World War have made Western Europe largely self-sufficient in food and a net exporter of some commodities. Grain production has nearly tripled due to improved varieties, pest control and near optimal use of fertilizers.

Current agriculture in Western Europe is an example of what can be achieved with a deliberate longterm policy of using science, finance and technology to increase agricultural output and thereby achieve self-sufficiency and food security. Present European agriculture is characterized by the following features:

- extensive mechanisation
- use of fertilizers to economically optimal levels
- regular use of plant protection chemicals
- use of high-yielding plant varieties
- use of livestock with emphasis on high productivity
- low input of labour.

This has led to a European agriculture which now produces food in abundance, in contrast with large shortages just after the Second World War.

Four aspects are important in describing recent developments:

- the formulation and introduction of agricultural policies related to production, 1945 - 1980
- the role of technology
- the effects of policies and technology on farming methods, systems and output
- the consequent position of European agriculture after 1980.

For the purposes of this chapter, Western Europe is taken to be the twelve present member states of the EC together with Norway and Sweden.

Agricultural policy 1945 - 1980

The present position of agriculture in Western Europe is largely a consequence of events in the Second World War.

The agricultural scene in Western Europe in 1945 was very different from that of today. World prices for agricultural commodities were relatively high, and home crop yields were less than half those of 1988. The farms were overmanned and underequipped. While advancing technology had led to the jet aircraft, the main source of power on farms was the horse. At a time when policy makers had the will to increase agricultural production, the technological means had become available to ensure their policies were effective. But there have been large differences between countries, both in the time it took to overcome the destructive effects of the war and resume agricultural development, and in the degree of progress since achieved.

General early objectives were:

- to ensure food supplies at acceptable prices to the consumer
- to provide reasonable levels of agricultural income so that rural areas remained populated.

The main methods adopted to achieve these objectives were price guarantees and import regulations, capital grants, favourable loans, free or subsidized advisory services, and land use and property restructuring policies.

By 1955, world agricultural output had improved relative to consumption, and prices had fallen. Emphasis in Western Europe began to move from increased domestic production to greater agricultural efficiency and stabilization of governments' expenditure. The need for a larger, supranational organization of agricultural markets was recognized. This need forms part of the basis for the Treaty of Rome and the EC. The political objective is (article 39 of the Treaty):

- to increase agricultural productivity by promoting technical progress and by ensuring the rational development of agricultural production and the optimum utilization of the factors of production, in particular labour
- to ensure a fair standard of living for the agricultural community, in particular by increasing the individual earnings of persons engaged in agriculture
- to stabilize markets
- to assure the availability of supplies
- to ensure that supplies reach consumers at reasonable prices.

Trends in Western European agriculture 1950 - 1989

Technology

Research and development in agriculture and related areas made significant advances in the first part of the twentieth century - development of industrial nitrogen fixation, introduction of the first herbicides and improvement of the internal combustion engine. The Second World War stimulated further developments. After the war, modernization of farming methods was supported by European governments with an expansion of state sponsored research and establishment of extension services. These programmes had broad public support.

Adoption of new technologies enabled a steady increase in farm productivity. Chemical pest control developed rapidly. Fertilizer production expanded with rapid technological advances. The greater use of fertilizer and emphasis on nitrogen were made possible by the introduction of new crop varieties with high yield potential.

In 1950 about 75 per cent of farm tractive power (EC9) was from animals. By 1975 this had declined to 5 per cent and the number of animals from about 10 million to 1.5 million, releasing large areas for other use (e.g., a horse needs 1 ha land for feed). Rapid mechanization and widespread rural electrification greatly increased labour productivity on farms and made possible the transfer of labour from farms to the cities. It also made agriculture dependent on external sources of energy.

There is therefore no single dominant technological factor in the great changes seen on farms. Instead, diverse technologies interacted and together enabled the large increase in farm productivity of the past forty years.

Farm structure

The increase in productivity made possible by new technology with the concomitant benefits of economies of scale, and the pressure of low produce prices have led to a rapid fall in the number of farms, particularly small holdings, Table 3.1.1.

As the total area of agricultural land has remained relatively static, the reduction in numbers of farms implies an increase in the average size of holdings.

Table 3.1.1 FARM HOLDINGS BY SIZE

Holdings (in thousands) (EC9)

Size	1960	1985	% changes
1 - 20 ha	6201	3169	- 49
20 - 50 "	813	799	- 2
Over 50 "	256	366	+ 43
	7270	4331	- 40

Source: Commission of European Communities Statistics.

The number of farms over 50 ha has increased, but these still account for only 8 per cent of all farms. At the same time there has been a tendency towards specialization, leading in some countries to the gradual geographical separation of the livestock and arable sectors and to a reduction in mixed cropping.

The family farm is still the principal unit in Western European agriculture.
A farm in Vestfold, Norway.
Photo: T. S. Knudsen.

Fertilizer

Use of inputs such as new varieties, pesticides and in particular fertilizer, increased rapidly during the period and resulted in substantial increases in output per unit of land.

Figure 3.1.1 WESTERN EUROPEAN FERTILIZER CONSUMPTION EC 12 + NORWAY AND SWEDEN

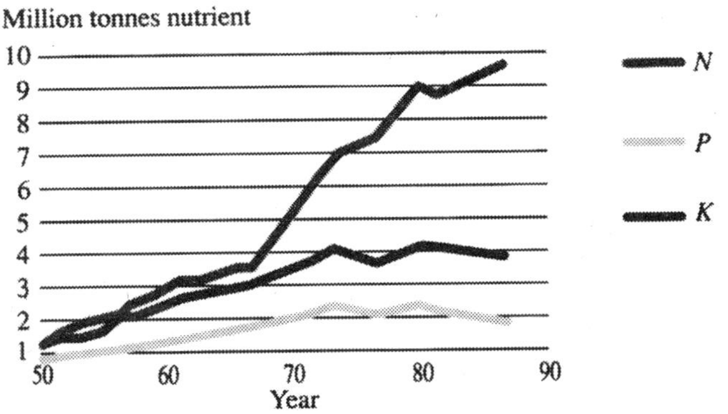

The trend in fertilizer consumption, Figure 3.1.1, shows that since the mid 1960s, most of the growth has been in nitrogen, but also suggests that consumption is now levelling off. Use of phosphorus and potassium increased steadily between 1950 and 1970 as these nutrients were replenished in impoverished soil. Since 1970, use has stabilised as applications are now made mainly to maintain, rather than to increase, levels of these nutrients in the soil.

Crop areas and livestock population

The total area of agricultural land has remained fairly constant at about 82 million ha arable land, and 58 million ha permanent grassland. Western Europe has no virgin lands and the loss of agricultural area to urban or road construction has generally been balanced by more intensive cultivation of poorer farm land.

The areas devoted to individual crops have, however, changed; in some cases markedly. Barley has expanded, while oats have declined. The total cereal area has, however, remained constant at about 38 million ha. The area of potatoes has declined sharply (from 4.3 to 1.5 million ha) but sugar beet, and oilseed rape in particular, have increased from a combined area of

Figure 3.1.2 WESTERN EUROPEAN LIVESTOCK POPULATIONS EC 12 + NORWAY AND SWEDEN

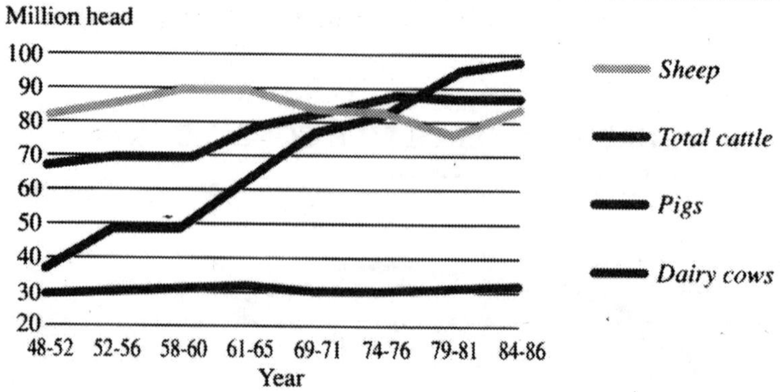

1.3 to 3.4 million ha. These changes in crop areas can be related to varying levels of subsidies and support provided by national governments or the EC. In livestock there have been increases in cattle and notably in pig populations, Figure 3.1.2. The horse population has declined dramatically.

Crop and livestocks yields and production

The main characteristic of the period is the large increase in both yield and production in virtually every agricultural sector. These are shown for cereals, Figures 3.1.3 and 3.1.4; for potatoes, sugar beet and oilseed rape, Figures 3.1.5 and 3.1.6; for milk, Figures 3.1.7 and 3.1.8; and for meat, Figure 3.1.9.

Figure 3.1.3 TRENDS IN CEREAL YIELDS EC 12 + NORWAY AND SWEDEN

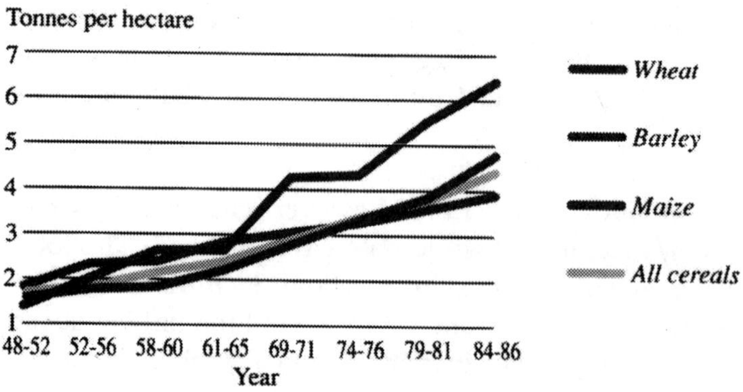

Figure 3.1.4 TRENDS IN CEREAL PRODUCTION EC 12 + NORWAY AND SWEDEN

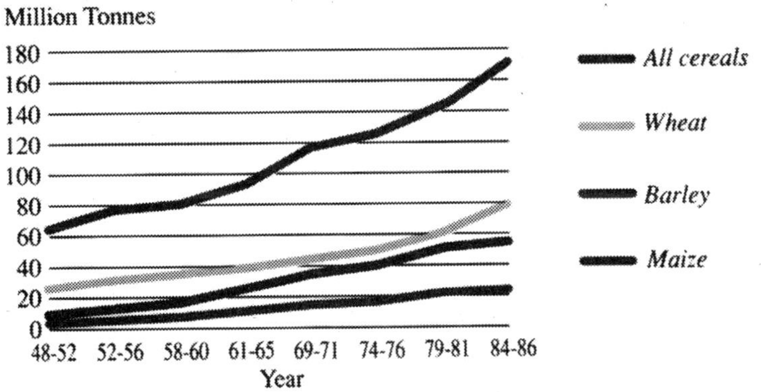

Figure 3.1.5 TRENDS IN POTATO, SUGAR BEET AND OILSEED RAPE YIELDS EC 12 + NORWAY AND SWEDEN

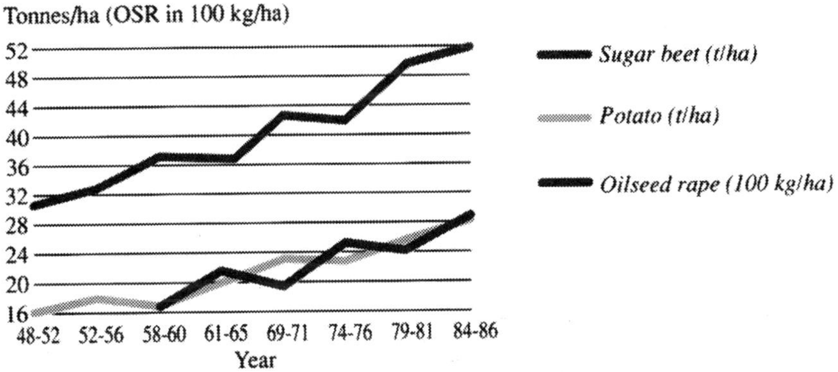

Figure 3.1.6 POTATO, SUGAR BEET AND OILSEED RAPE PRODUCTION EC 12 + NORWAY AND SWEDEN

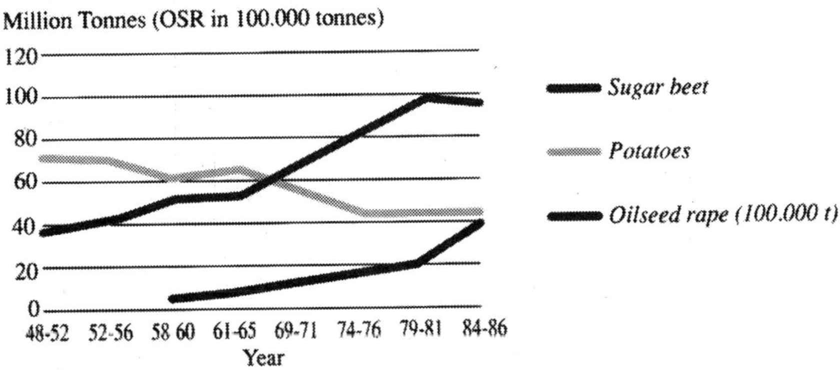

Figure 3.1.7 MILK PRODUCTION IN EC 12 + NORWAY AND SWEDEN

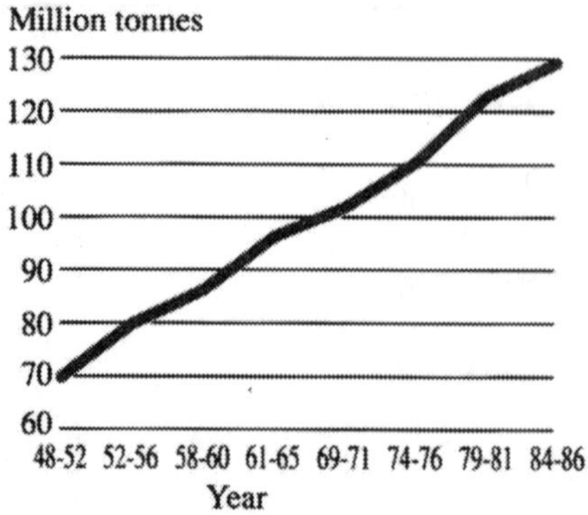

Figure 3.1.8 MILK YIELDS IN EC 12 + NORWAY AND SWEDEN

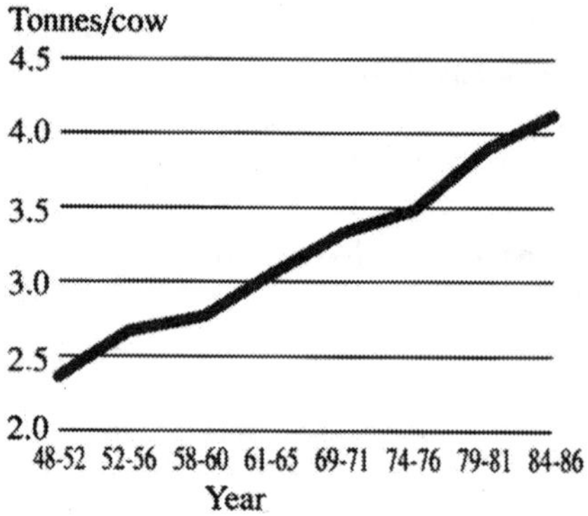

Figure 3.1.9 MEAT PRODUCTION EC 12 + NORWAY AND SWEDEN

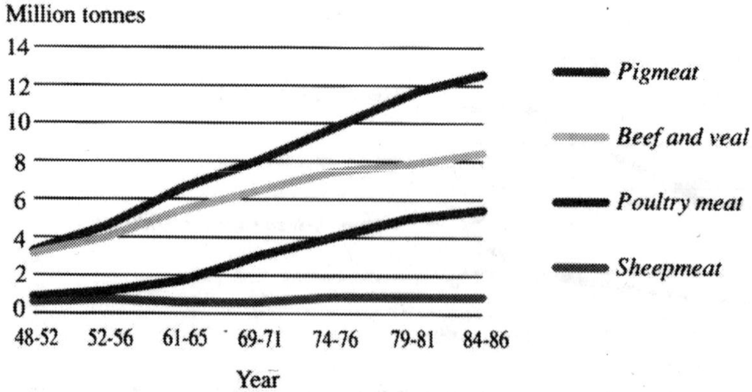

The European food market

Population

The main purpose of agriculture is to provide people with food. The total population of Western Europe grew from some 270 million to nearly 340 million between 1950 and 1986, which represents an annual (compound) growth rate of 0.6 per cent.

Although the growth in population has significantly increased food consumption, this growth has been much slower than the increase in agricultural output. The number of people working in agriculture, as a proportion of the total population, has declined from 10 per cent in 1950 to 3 per cent in 1986. However, this underestimates the total number of people dependent on agriculture for a living because a substantial number are employed in the supply industries and in food processing and distribution.

Self-sufficiency in food production

Reliable information on the levels of self-sufficiency is difficult to find for the early years. In many countries of Western Europe it will have been in the range 40 - 70 per cent in 1950. From the current debate on surpluses, it is clear that the degree of self-sufficiency rose in the intervening years. Trend data are available for the UK, Figure 3.1.10.

Figure 3.1.10 THE VALUE OF HOME PRODUCED FOOD AS A PERCENTAGE OF ALL INDIGENOUS TYPE FOOD CONSUMED IN THE UK

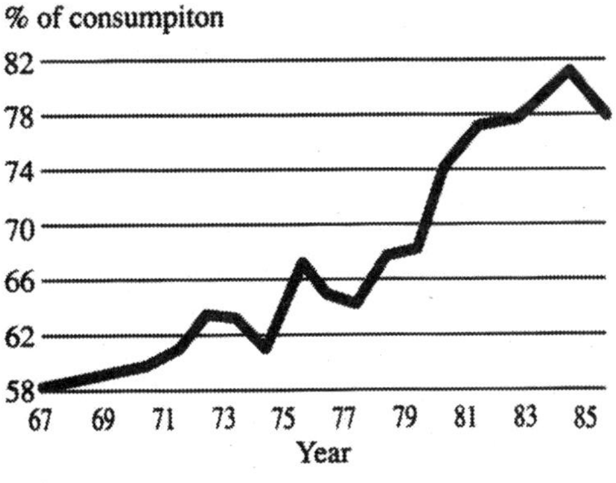

Numbers for the EC10 are given in Table 3.1.2. These EC figures conceal a wide variation between nations in the degree of self-sufficiency largely related to specialisation of production. For example, in France, total self-sufficiency in cereals in 1985/86 stands at 204 per cent while the correspon-ding figure for the Netherlands is 25 per cent.

Table 3.1.2 SELF-SUFFICIENCY (%) IN CROP PRODUCTS (EC10)

Crop product	1973/74	1985/86
Total cereals	91	119
Potatoes	101	102
Fresh vegetables	94	101
Sugar	NA	133
Whole-milk powder	231	342
Butter	98	117
Beef	95	108
Sheep meat	66	76
Poultry meat	102	105
Pig meat	100	102

Source: Commission of European Communities Statistics.

The general rise in self-sufficiency is a consequence of stable demand (small population growth) and greatly increased agricultural productivity.

The position of agriculture after 1980

Changes in agricultural policies can be expected and have occurred as food production approaches, then exceeds, self-sufficiency. Notable exceptions are for crop products still in deficit (e.g., oilseeds, protein crops) and for countries where general self-sufficiency remains low (e.g., Norway).

Further, while Western Europe now benefits from the abundance and variety of food, there is increasing concern about the environmental consequences of the policies that have been pursued.

This, together with growing official concern about the costs of agricultural support, particularly in the EC, has resulted since 1980 in new political initiatives in agriculture. These have been aimed at:

- curbing the production of commodities in surplus
- import displacement
- reducing the cost of agricultural support
- reducing the environmental impact of agriculture.

Some recent key events have been:

1982/84 - Sweden introduces fertilizer taxes to contribute to environmental research and advice

1984 - EC introduces milk quotas

1987 - EC introduces Environmentally Sensitive Areas scheme (farmers paid to manage land in an approved way)

1987 - EC changes and reduces price support for cereal production

1987 - Baden-Württemberg: Fertilizer nitrogen use in water production areas is limited to 20 per cent below optimal level

1988 - CAP market expenditure limited to real growth of less than 2 per cent per year in the period to 1992

1988 - EC introduces area set-aside. UK, West Germany and the Netherlands set up schemes

1988 - UK abolishes all farm capital grants related to production. These are replaced by grants for conservation

1988 - In the Uruguay Round of GATT, USA calls for the complete elimination within a decade of all farm and export subsidies in agriculture

1989 - Proposed EC directive to apply the 50 mg nitrate/l limit for drinking water to water sources. This allows for controls on agriculture including manure and fertilizer nitrogen use.

Other regulations could be mentioned. The list illustrates the present trend and it is not a complete catalogue of actions or legislation in this area.

These events suggest that agriculture in Western Europe is about to develop in new directions as new objectives are defined and the economics of farming are adjusted to ensure that they are achieved. The farmer may find himself with increased responsibilities and opportunities as a countryside manager. Production of food, industrial crops and forestry products will remain the principal objective, but farmers will increasingly supplement their incomes by assuming duties related to:

- wildlife protection and promotion
- landscape conservation and maintenance
- provision of services and facilities for leisure and education.

References: Burrell et al. (1984, 1987), Commission of European Communities (1978, 1985a,b, 1988), FAO (1950-1958, 1959-1986, 1960-1986), HMSO (1948-1988).

Rural nature requires care and maintenance. From Hardanger, Norway. Photo: T. S. Knudsen.

3.2 Current farming methods in Western Europe

A wide variety of agricultural systems can be found today in Western Europe. In some parts, farming methods and the appearance of the countryside have changed beyond recognition in the past forty years. However, farms can be found that would be familiar to an inhabitant of the last century.

Crops grown vary from olives and vines to cereals, potatoes and oilseed rape. The area of grassland includes mountain meadows and intensively managed grass for silage. Nevertheless, there are some characteristics that are common to the majority of European farms - growing reliance on science and technology being the most obvious.

Despite the diversity, these farming systems have enough in common to justify use of the term "current agriculture" for farming conforming to good agricultural practices according to local conditions, with proper use of fertilizers and pesticides (Zadoks (1989)). Conventional agriculture is another name in common use. For a description of current farming in France, see Soltner (1988a,b).

A typical West European farmer today has a clear financial objective for his enterprise, and generally will manage his farm as a business. Financial advice will be available to him, very often subsidised or offered free by his national government.

The main cost elements of arable farming are (as parts of total costs):

- fixed costs (buildings, machinery, interest, labour, etc.) — about 50 - 70 per cent
- variable costs:
  - fertilizers — " 10 - 20 " "
  - seeds, sprays, etc. — " 15 - 35 " "

The farmer's net income is the difference between costs and income from produce sales. It thus depends critically on yields. Reduced inputs usually give a yield reduction that is not fully compensated by reduced variable costs.

Until recently, most farmers have been protected from commodity market forces by guaranteed markets for their produce. This protection is now beginning to weaken and farmers are becoming more sophisticated in the marketing of their output - as vegetable growers always have been.

Farming methods are now applied consciously to achieve the farmer's financial objectives. Fertilizers and agrochemicals can together account for around 80 per cent of the variable costs in cereal production and their use must be controlled if margins are to be maximised. The farmer cannot afford wasteful use of these inputs.

The amount of fertilizer used is normally based on recommendations available from government agencies or manufacturers/distributors. Crops differ in their requirements and recommendations also reflect local conditions, but will often be in the range:

- Nitrogen (N) 100 - 250 kg/ha
- Phosphorus (P) 20 - 35 "
- Potassium (K) 40 - 170 "

Fertilizer nitrogen is normally given in one or more applications during the crop or grass growing season when it can be taken up rapidly and become effective in promoting yield. Phosphate and potassium are less subject to the risk of loss; spring and autumn applications are most common.

Agrochemicals and pesticides differ from fertilizers in that their active ingredients are not found naturally in the soil, while those of fertilizers are. They are applied to the seedbed or the growing crop to control specific weeds, diseases or pests. Prophylactic use of agrochemicals is now discouraged and controls over application are likely to become steadily more restrictive.

The use of fertilizers and agrochemicals has partly removed the need for fixed arable or arable/grass rotations. These crop rotations were designed to prevent depletion of the soil's nutrient reserves, to replenish nitrogen by growth of a legume and to prevent the build-up of weeds, pests and diseases each of which tends to be associated with a particular type of crop.

The introduction of fertilizers and agrochemicals now allows more flexibility in cropping patterns and the production of foods in proportions that match demand. The move away from rotation has also been promoted by developments in farm machinery. As equipment has become larger and more expensive, greater areas of a crop have been needed for convenience of operation and to reduce costs. The main field mechanical operations on farms are soil cultivation, application of fertilizers and agrochemicals and harvesting. Of these, cultivation is especially important in relation to nutrient cycling and losses.

Soil cultivation is done for two main reasons - to reduce soil compaction in order to allow seedling roots to develop, and to control weeds. The objective is to break up the surface layer of soil typically to a depth of up to 30 cms and, where a plough is used, to invert this soil layer (so burying weeds). Any cultivation results in mixing soil and air which promotes the mineralization of organic matter and the release of nitrate.

Increased mechanisation and the move away from crop rotation have allowed a reduction in farm labour and the modern farm employs few people. Farm workers have become more skilled than their predecessors, especially in the use of complex machines such as combine harvesters and mechanised handling systems. Current farming allows the full productive potential of land to be achieved. Crops are relatively free from disease and nutrient deficiencies, and are of a consistent quality.

The livestock sector is now quite different from that of forty years ago. Grasslands are managed more intensively and in many areas are treated much as an arable crop with optimal levels of fertilizer use. In some regions, there is a concentration of intensive animal production units based on housed animals and poultry, with an output of manure in excess of what should be applied on local crops.

Maximum livestock output on a farm can only be achieved if the animals are healthy and relatively free from stress, but the living conditions can appear cramped and "unnatural" and are a cause of current public debate.

Veterinary care includes the use of drugs to control disease. Recently hormone treatments have become available to increase growth rate and increase milk production, but the use of these measures is likely to be severely controlled.

The modern European farm is operated to ensure efficient use of physical resources so as to provide a positive financial return to the farmer. Political decisions, economics, science and technology are the driving forces.

Ploughing in autumn with a modern reversible plough.
Photo: Norwegian Agricultural and Marketing Cooperative.

Animal husbandry forms a principal part of European agriculture. Dairy herd in Denmark.
Photo: Norsk Hydro, Denmark.

CHAPTER 4

ENVIRONMENTAL ISSUES RELATING TO FERTILIZER USE

4.1 Introduction: a listing of the environmental issues relating to fertilizer use

Environmental issues relating to farming and farming practices have come to the fore in recent years. Some of the controversies concern fertilizer use, others do not.

The subject of this chapter is environmental and health issues relating to fertilizer use.

These issues are:

Section 4.2: fertilizers and soil quality:

- fertility
- acidity
- structure
- nutrient reserves and balances
- organic matter
- biology: microbes and earthworms.

Section 4.3: nitrogen in agriculture:

- the low overall efficiency of nitrogen input
- ammonia emissions
- nitrate leaching and nitrate in drinking water
- nitrate toxicology

Section 4.4: phosphorus:

- phosphate losses to water

Section 4.5: the remaining nutrients:

- potassium leaching
- sulphate in water
- mineral nutrient imbalances in herbages
- soil nutrient exhaustion.

Section 4.6: fertilizers and the other elements:
- cadmium
- arsenic
- fluorine
- aluminium
- radioactive elements

Section 4.7: erosion and fertilizer use

Section 4.8: eutrophication of fresh and marine waters

Section 4.9: fertilizers and crop pests

Section 4.10: fertilizers and produce quality

Section 4.11: energy use in agriculture

Section 4.12: depletion of mineral resources

Section 4.13: pollution from fertilizer manufacture

Section 4.14: indirect environmental effects of fertilizer use:
- reduction of the diversity of the natural environment
- issues concerning specialized mechanized agriculture.

The importance and relevance of these issues vary among regions, depending on soil, climatic and other factors. Discussions in this chapter mainly refer to conditions similar to those in Europe, though the principles outlined should also apply elsewhere.

References: Popular books on environmental issues of food production and fertilizer use are available in various languages:

Danish: Aslyng (1978);
English: CEA, IFA, IPI (1983), Jollans (1985);
German: BASF (1985), Vetter (1980), Wiedemann-Sander (1987);
Swedish: Royal Swedish Academy of Forestry and Agriculture (1987).

.2 Soil quality

The fertile soil

Plants can grow in water culture without soil (hydroponic horticulture) but almost all crops are grown in soil. The soil provides:

- mechanical support
- space for root growth and development
- air (oxygen) for root respiration
- water
- nutrients
- a medium for interaction with other organisms, both beneficial and harmful.

The growth and development of plants suffer where soil conditions are detrimental to one or more of these functions.

Soil originates through the disintegration of parent rock material, a process known as weathering. Soils develop through interplay between physical (e.g., frost, grinding) and chemical processes such as the action of acids, dissolution and precipitation. Plants and soil microbes and animals gradually form this material into soil through their growth, decomposition and residues.

Mineral particles, live and dead organic matter, air and water are the main constituents of soils. Soils show a wide variety of chemical, physical and biological properties depending on their origin and history. Soil properties are not static; they change through time and are influenced by cropping and cultivation practices. Various systems are in use for soil classification as guidance for their agronomic properties. These are described in standard textbooks on this fundamental subject (e.g., Wild (1988), Scheffer and Schachtschabel (1976)) but the following concepts form a background to the rest of this chapter.

Soil particles are classified according to size:

- sand (2 - 0.05 mm in diameter)
- silt (0.05 - 0.002 mm)
- clay (less than 0.002 mm).

Sand and most silts are chemically inactive. The chemical properties of the soil depend mainly on the nature of the clay particles, soil organic matter and their capacity for controlling soil acidity and binding and releasing nutrients and noxious components.

Soil *texture* depends on the relative proportion of sand, silt, clay and organic matter. It can be approximately determined by the sense of feel in the field. A "light" or coarsely textured soil is sandy while a "heavy", fine textured soil has a high clay content. Loam is a rather loose soil of silt, clay and organic matter. A soil with more than 40 per cent organic matter is usually termed "organic" (e.g., peat soils). Texture is a guide to the ease of cultivation of a soil. Light soils are relatively easy to cultivate while heavy soils are more difficult. On heavy soils, several operations are usually needed to form a seedbed and often a delay is necessary between ploughing and sowing to allow weathering of the large soil clumps. Heavy soils are also difficult to plough when wet over winter. For these reasons, and because there is only a short period in which spring crops can be sown, heavy soils are often ploughed in the autumn.

Soil *structure* describes the aggregation of soil particles, Figure 4.2.1.

Figure 4.2.1 SOIL STRUCTURE (FAO (1984))

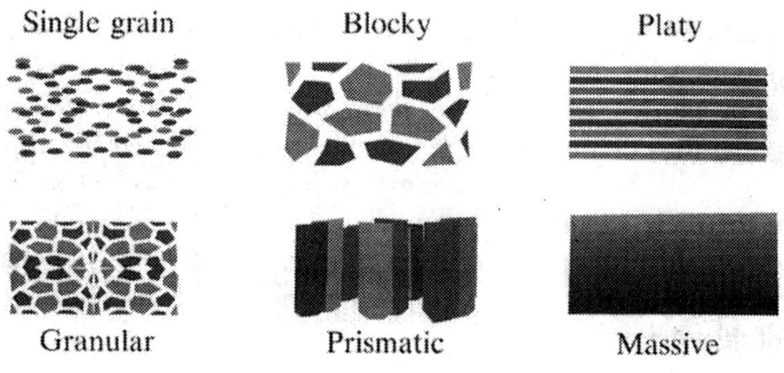

Aggregation depends on soil texture, organic matter content and chemical factors. Soil porosity is mainly related to the size of aggregates. Water and air movement in the soil and ease of root growth depend on soil porosity and thus on structure.

Three examples of different soils are shown below.

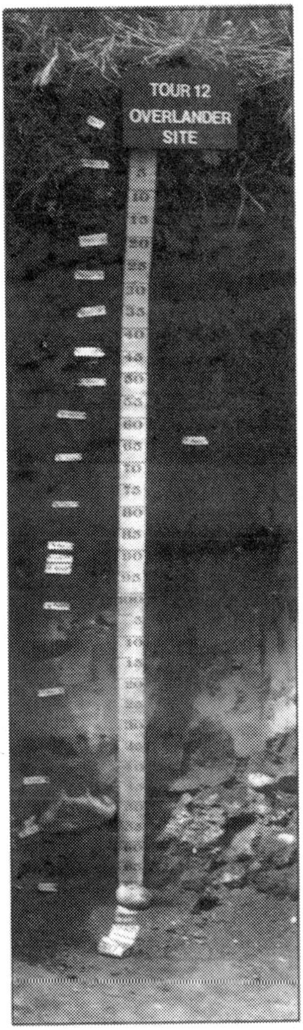

The chernozems or black earths are the world's most fertile soils. This picture from Alberta, Canada shows the layering and the influence of plants in soil formation.

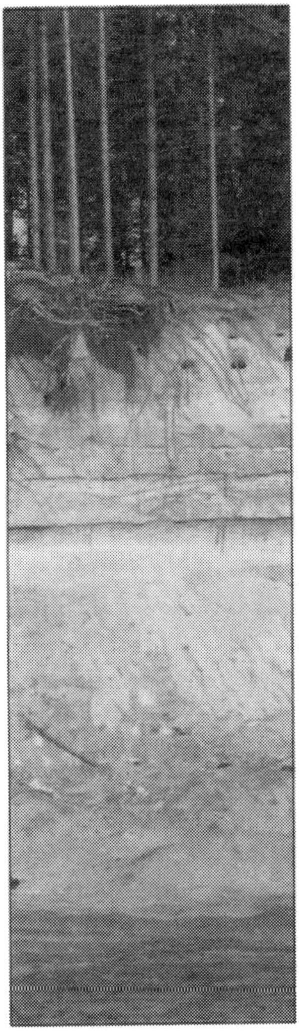

A typical sandy soil poor in nutrients, from Brunlanes, Norway. Such soils are productive when they are properly fertilized.

A typical heavy clay from Hyderabad, India. This soil forms deep cracks and large aggregates during a drought.

Photos: O. Ø. Hvatum, Department of Soil Sciences, Agricultural University of Norway.

A soil with satisfactory porosity, moisture retention, aeration and drainage characteristics is said to have a good structure. Improvement of the soil structure is one of the purposes of soil cultivation.

During soil development, fine particles and dissolved matter follow downward water movement and can precipitate in deeper layers of the soil. Thus layers with different colour, texture and properties develop. In an arable soil, the top layer is mixed through cultivation, and a topsoil develops to plough depth (about 20 - 30 cm). Though this is the main region for root development, the roots of most crop plants can grow down to about 1.5 m and utilize water and nutrients in these deeper soil levels. Some crops (e.g., potatoes) have a shallower root system, down to about 1 m.

Soil fertility

The term soil fertility, especially in older literature, is used to describe a soil's ability to support crops through its own nutrient reserves. We will use it in this sense although there is no generally accepted scientific definition. Soil fertility is a loosely used term and different groups use it in different senses.

Fertile soils are formed under favourable climatic conditions from a parent material rich in mineral nutrients, with nitrogen reserves originating mainly from biological nitrogen fixation. Soils like these may give good crops for a considerable time before the level of one or more nutrients becomes inadequate. This can give the impression that such soils are adequate and necessary for a sustainable agriculture, and that large soil nutrient reserves are a prerequisite for high yields.

The related term "soil productivity" is used to describe a soil's ability to support crops when cultivated correctly, including supplements of organic and mineral nutrients to compensate for removal and losses and to maintain soil reserves.

Soils which are poor in nutrient reserves (e.g., light sandy soils) can be highly productive when fertilized and managed properly. Their small capacity to retain nutrients and water means that cultivation techniques designed to control nutrient leakage are essential (Brüne (1988)).

The difference in fertility between such light soils and those traditionally regarded as fertile is one of degree rather than principle; all soils sooner or later become exhausted through long term continuous cropping if the nutrients that are removed are not replenished. This had happened to some of Europe's arable land before fertilizers became available.

Soil organic matter is often described as the key factor in soil fertility, but it is only one of a number of factors which determine the suitability of soils for crop production. Fertility is also influenced by soil type, the nature of any clays present, water status and biological activity. Productivity of the soil is further determined by agronomic factors including disease control, nutrient supply, crop variety and past management.

When cultivating a soil, the farmer strives to improve soil productivity through a variety of interdependent measures. Emphasis on one factor alone is usually not enough:

- fertilization without adequate water supply may prove worthless
- all the nutrients must be given attention. Thus the same nitrogen application gives better yields on soils with a good phosphorus and potassium status than on those with a low supply of either
- soil conditions should be maintained or improved, e.g., by appropriate tillage, liming and return of crop residues to the soil.

References: Foth and Ellis (1988), Marschner (1986), Tisdale et al. (1985), Wild (1988).

Yields from natural fertility

Some soils have a high natural fertility; these are usually mineral soils with a high humus content. In the short term these soils can give crop yields up to 75 per cent of those obtained on properly fertilized land. But over a long period of time, no soil can continue to give high yields without replacement of removed nutrients. If a large proportion of the nutrients is recycled, this will slow down depletion, but since some nutrients are exported out of the system and some are otherwise lost, the fertility of the soil will gradually decrease. The need to prevent this steady depletion, especially of phosphorus and potassium, is the basis for fertilizer applications to many crops; the amount of the nutrient removed in a crop is replaced by a subsequent fertilizer application so ensuring that the nutrient status of the soil is maintained.

Generally, long term experiments show that the yield level obtained without any fertilization is of the order of 25 to 45 per cent of the optimum yield under current agriculture. This is what natural fertility can support. In such systems the removal of nitrogen, phosphorus and potassium will be about 15-35, 5-15 and 10-25 kg/ha • year, respectively. Nitrogen is supplied through nitrogen fixation by legumes, algae and bacteria and by precipitation, while phosphate and potash are supplied by weathering of soil particles. Circulation on the farm of nutrients in organic materials like manure, compost, etc., is not a net input unless animal feeds (e.g., concentrates) are brought in from outside.

Figure 4.2.2 EFFECT OF FERTILIZERS ON YIELD

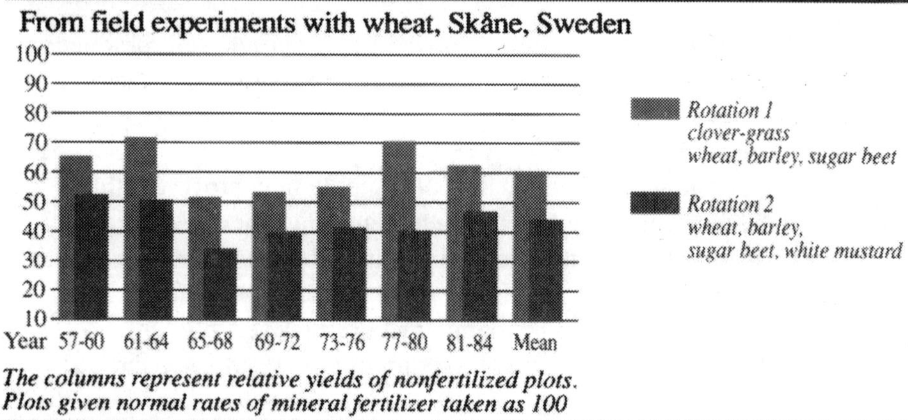

Source: Nilsson (1986).

Figure 4.2.2 shows the results of a long-term experiment in Sweden comparing two different types of management and crop rotation, with and without fertilizers (Nilsson (1986)). Both had a four-year rotation of wheat, barley and sugar beet. Rotation 1 had in addition one year of a clover-grass ley. The clover-grass and sugar beet tops were harvested in rotation 1 and corresponding amounts of manure returned to the land. Rotation 2 included an oil seed crop (white mustard) in the fourth year of rotation, and all plant residues were ploughed down.

Yields on plots given no mineral fertilizers were compared with those where normal rates were used, the yield on plots given mineral fertilizer being taken as 100. The blue and red columns represent the relative yields of wheat in the two rotations on plots receiving no fertilization.

Rotation 1 carried commercial crops in three out of four years. In order to get comparable yields on the arable part of a farm, yields in rotation 1 should be correspondingly reduced from those given in Figure 4.2.2. In doing so, it is found that the commercial yield of rotation 1 without fertilization is 47 per cent of the yield in rotation 2 with normal rates of NPK-fertilization. This would be the relative output of a mixed farming system without mineral fertilization compared with specialized arable cropping on this place and soil.

Soil physical and chemical factors

Soil acidity

Acidification of soils is a natural process, caused by the formation of inorganic and organic acids through microbial activity, and by loss of soil bases through ion exchange and leaching with excess rain water. Root uptake of cations (e.g., K^+, Ca^{2+}) is accompanied by excretion of acid (H^+) and thus contributes to soil acidity. The extent of this depends on plant species and growth, legumes especially having an acidifying effect. Natural acidification of soil is now enhanced by the acids in rain and dust originating from an industrialized society.

Soil acidity is a serious problem in some areas, especially in forests, but in agriculture, acidity is routinely corrected by liming.

Soil acidity can be measured as the acidity (pH) of water in equilibrium with the soil. It ranges for mineral soils between about 3.6 and 9.0. Values between 5.5 and 7.5 are most common for agricultural land.

Microbial action in soil converts ammonium to nitric acid. Any input or process that increases the supply of ammonium will tend to promote soil acidity. The nitrogen fertilizers most used today generally have an acidifying effect, mainly due to the ammonium content, Table 4.2.1. Calcium nitrate is an exception. Its ammonium content is low, and the calcium content gives it some liming value, as indicated in the table.

It is not acidity as such that damages plant roots. Plants can grow in soilless solutions at a pH below 4.5. But in acid soils with a pH below 5.5 - 5.0, the solubility of aluminium and manganese increases. These metals in ionic form can damage the roots where soil pH is low.

There may be yield reductions on acid soils without any visible damage to the plant. This can be caused by less favourable conditions for microbial activity, stronger binding of phosphate to aluminium and iron, thus making it unavailable to plants, and less favourable physical conditions in the soil. If the pH is high, above 7.5, iron, manganese, zinc, copper and boron will be strongly bound in the soil. This can cause deficiencies and reduced crop yield. Phosphate is most readily available between pH 5.5 and 7.5. It is important that soil pH is adjusted to the range appropriate for the soil and crop through liming.

Table 4.2.1 EFFECT OF FERTILIZERS ON SOIL ACIDITY

The amount of lime (as CaO) required to neutralize the acid generated from the application of 100 kg N with the fertilizer specified.

Type of fertilizer	Per cent N as Nitrate	Per cent N as Ammonium	Kg CaO per 100 kg N		
Calcium ammonium nitrate	13.0	13.0	35	to	75
Ammonium nitrate	16.5	16.5	80	"	120
Ammonium sulphate		21.0	240	"	320
Liquid ammonia		82.0	40	"	120
Urea			40	"	120
Nitrophosphate	9.5	11.5	85	"	130
Calcium nitrate	14.5	1.0	-50	"	-100

Liming (the addition of ground chalk or limestone or related materials) is a standard measure to improve the soil. Liming is done to:

- adjust soil pH and counteract soil acidification
- maintain levels of available calcium in the soil
- improve the structure of heavy clay soils
- improve the supply of some nutrients
- promote microbial activity and thus mineralization of the soil's organic matter
- prevent damage to roots from soluble aluminium.

Reference: Robson (1989).

Nutrient interactions

The plant root is able within limits to regulate its uptake of nutrients. The balance between nutrients in soil solution also influences their uptake, e.g.,

- a high concentration of potassium can reduce magnesium uptake by crops
- a high concentration of sulphate reduces molybdenum uptake.

Fertilizer plans and recommendations are best made for individual fields on the basis of soil or crop analysis or other methods. This enables the removal and addition of nutrients to be balanced, so avoiding unnecessary application or soil exhaustion.

Structure

Soil structure is constantly changing under the influence of the mechanical forces and water movement that result from rain, evaporation, freezing, thawing and water uptake by plant roots. Biological processes such as root exploration, earthworm burrowing and microbial transformation of organic residues also exert a major influence.

The heavy equipment used today for cultivation and harvest can damage soils (Kuipers 1982) through:

- compaction: reduction of soil porosity through compression
- smearing: destruction of aggregates and mixing of their components resulting in the formation of a hard layer of soil.

In many soils a compacted hard pan is formed just below the plough layer. Such a pan can be very difficult for both roots and water to penetrate.

The extent to which these processes can occur varies with the soil's water content. Drainage and timing of operations in relation to weather are examples of soil protection measures.

Soils damaged by compaction or smearing can be regenerated by natural processes, e.g., drying, freezing and the activity of soil animals but such regeneration can be a very slow process.

Aggregate stability is improved by humus, hence the positive effect of farmyard manure on soil structure. Mineral soils are often improved by timely inputs of organic material. But since other factors, e.g., liming, freezing and drying, also help to form aggregates, organic matter can be fairly low without detrimental effects on structure.

Fertilizers are often cited as having a detrimental effect on soil structure, mainly because they are thought to cause a decrease in the soil's content of organic matter. But fertilizers do not *per se* have this effect.

Soil organic matter

Soil organic matter originates in living organisms. It is a factor of importance in soil productivity because:

- nutrients are bound to soil organic matter and released on its decomposition
- some of it acts as food to soil organisms
- it stabilizes mineral soil aggregates.

The biological cycle

Organisms eventually die and decompose, thus returning essential elements to the soil to be taken up again in living matter. This circulation is called the biological cycle.

The formation phase is characterized by:

- photosynthesis and the formation of organic matter
- the uptake of mineral nutrients by plants.

The decomposition phase is characterized by:

- transformation through consumption into living matter in other organisms
- step-by-step degradation of complex organic molecules into simple ones
- liberation of inorganic compounds (mineralization)
- transformation of organic matter into a rather stable substance called humus.

As can be seen from Figure 4.2.3, plants have a central place in the formation phase of the biological cycle, and synthesise organic compounds from mineral nutrients, carbon dioxide and water. Decomposition is mediated by bacteria, fungi and soil animals.

Organic matter in the soil consists of:

- living plant roots, bacteria, fungi, animals (discussed later in this section)
- exudates from plant roots and soil organisms
- dead plants and other organisms in various stages of decay, ranging from recent roots to stable humus.

Figure 4.2.3 THE BIOLOGICAL CYCLE

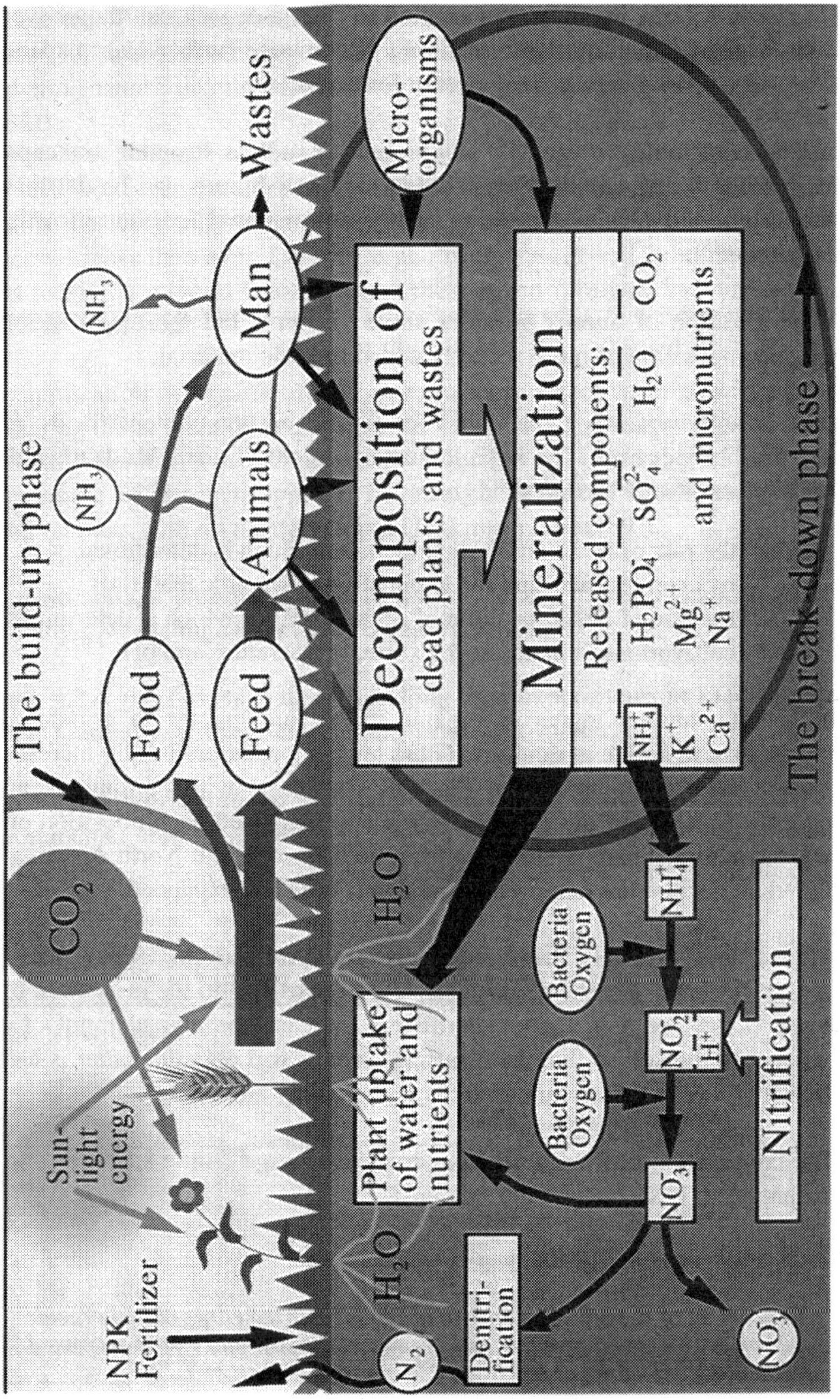

Humus is organic residues transformed to such a degree that they have lost their original structure. Some of it may decompose further over a span of a few years, other fractions may persist for centuries.

Humus contributes to valuable soil properties such as structure and capacity for holding water and nutrients. Large losses of humus can be detrimental to soil productivity, but humus as such is not essential for plant growth and development.

Mineralization of humus provides some nutrients but the major source is the decomposition of more recently added organic material.

Mineral soils typically have values for organic carbon content* in the range of 0.5 to 2.5 per cent. The organic matter content of a soil tends towards an equilibrium level which depends on:

- the rate of addition of organic matter which is determined by crop residues and the application of organic materials
- the rate of decomposition of organic matter which is determined by cultivations, climate, soil texture, temperature and pH.

In general, annual inputs of organic matter are greater in grasslands or forests than in arable agriculture. Grass leys in a rotation usually increase the amount of organic matter in the soil compared with continuous arable systems. Cultivation of land usually leads to increased decomposition of soil organic matter. There is evidence from both Europe and North America that a gradual decline has occurred as the arable area has expanded.

After a time, typically measured in centuries, organic matter stabilizes at a new lower level. Fertilizers correctly used contribute to the protection of the levels of soil organic matter. Fertilizers increase the annual input of crop residues to the soil so that the resulting level of soil organic matter is usually greater where fertilizers are used than where they are not.

The continuous fertilizer trials that cover the longest time span are those at Rothamsted, UK.

* *The content of organic substance in a soil is reported either as soil organic carbon or soil organic matter. Soil organic matter content can be estimated from measured values of soil organic carbon multiplied by 1.72.*

In these trials winter wheat has now been grown continuously for nearly 150 years. The straw is removed. The plots that are compared are given either farmyard manure or mineral fertilizer only, or no nutrients at all (Johnston (1982)).

The levels of organic matter have remained unchanged for the last 100 years in plots receiving only mineral fertilizers and yields of wheat on these plots are now higher than ever. Despite large differences in soil structure between plots receiving mineral fertilizers and those given farmyard manure, similar yields are obtained.

The application of organic matter is even more effective in protecting and promoting the amount of organic matter in the soil. In the Rothamsted trials, farmyard manure was six times as effective as mineral fertilizer (per kilogramme of nitrogen applied) in raising soil humus levels above those found on plots with no nitrogen input (Jenkinson (1988/89)).

In arable soils, a combination of fertilizer use and manure application will generally give the highest level of organic matter.

Figure 4.2.4 gives average data from long term experiments at four different sites in Germany with different inputs of organic matter.

Experiments with fertilized and unfertilized barley in Sweden illustrate the importance of crop residues for the supply of the soil's organic matter, Table 4.2.2.

Figure 4.2.4 THE SOIL CONTENT OF HUMUS AFTER DIFFERENT TREATMENTS IN LONG TERM EXPERIMENTS

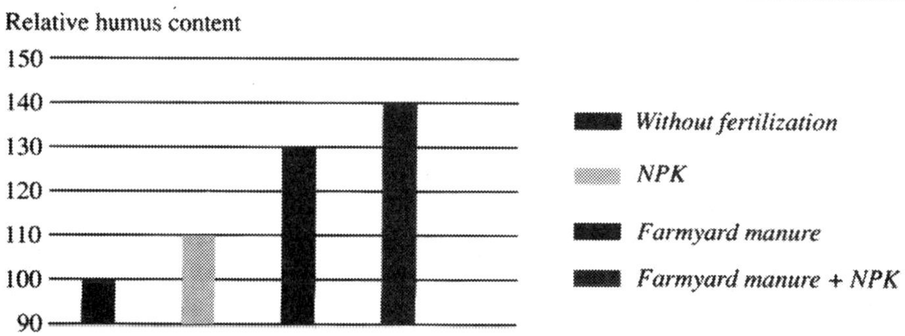

Mean of four experiments after 50 years. FRG.

Table 4.2.2 PRODUCTION OF DRY ORGANIC MATTER IN BARLEY WITH AND WITHOUT NITROGEN FERTILIZATION

Fertilization kg N/ha	Dry organic matter t/ha ear	straw	living roots
120	4.7	3.8	1.6
0	2.0	2.3	1.3

Source: Hansson et al. (1987).

The organic matter in roots (Table 4.2.2) underestimates the addition of organic matter by a crop to the soil as the figures exclude root exudates and the loss of parts of the roots during the season. Estimates of the proportion of the total assimilated carbon that is released into the soil by cereals range from 16 to 33 per cent. However, the amount of organic matter in the soil can in some circumstances be rather insensitive to differences in the amounts of crop residues returned to the soil (Jansson (1986)).

Topsoils are richer in organic matter than the deeper soil layers. Increasing plough depth temporarily reduces soil organic matter by dilution but the equilibrium level is gradually restored and the total amount of humus in the new, more deeply ploughed layer can even increase. This has happened over large areas in recent times.

In some cases excessive amounts of organic matter can accumulate, due to the lack of biological activity in acid soils, low temperatures or lack of aeration. Plant growth on such soils is usually reduced. This can be associated with severe climatic conditions, but frequently, it is due to the lack of nitrogen and other minerals. Thus organic matter *per se* is not a direct measure of soil fertility.

Nevertheless, there does appear to be a minimum level of organic matter (usually exceeded in European soils) below which crop growth and yield are adversely affected.

It is difficult but possible to improve soils with poor structure if the cause is a low level of organic matter. For example, a silty, fine-textured soil with a low level of organic matter and with poor structure could be put into herbage crops to build up organic matter and thus improve the structure. This implies sheep, dairy or cattle farming, or a viable market for the herbage.

Some clay soils that are low in organic matter may not be suited to continuous arable cropping. In these cases, grass leys can be used in a rotation to maintain an adequate soil structure. Other soils with a high level of organic matter and inadequate levels of nutrients, such as deep peats, are difficult to manage properly. Applying fertilizers and enhancing the breakdown of the organic matter by the addition of lime and by drainage and cultivation are usually employed, but care is required to prevent structural deterioration.

References: BASF (1985), Follett et al. (1987), Newbould (1982), Sauerbeck (1982).

Fertilizers and soil life

The processes of soil life are very important influences on soil formation and soil fertility (DLG (1988)).

Soil life is very diverse. It consists of micro and macro-organisms (bacteria, algae, fungi, animals such as protozoa, nematodes, earthworms and insects) and above all the plants themselves with their root systems. Their residues and root exudates form the principal source of nutrients for soil life.

Soil populations and their activity change with soil type and conditions (e.g., temperature and water supply). Increasing concentration in the soil of some heavy metals and pesticides can have adverse effects.

Micro-organisms mineralize soil organic matter and thereby liberate nutrients, but, developing microbial populations also incorporates nutrients into their living tissues. Humus is created and decomposed, pathogen and parasite populations are maintained or diminished. From the farmer's point of view, some aspects of soil life are beneficial while others are detrimental.

Plants and their sphere of influence

The dominating organisms in a soil are the plants themselves, through their roots. A special environment is created around the roots, through:

- uptake of nutrients and excretion of bicarbonate and other ions
- excretion of organic substances (mainly sugars and acids)
- loss and decomposition of parts of the root surface.

Plant roots have a pronounced influence on the surrounding soil. The photo shows a mixed culture of sorghum and the legume chickpea growing in the presence of a pH indicator. Nitrate is the nitrogen source. The bulk soil has a pH of 6.0 (brown) but the rhizosphere pH varies between 7.0 (purple, sorghum) and 4.5 (yellow, chickpea).

Photo: V. Römheld, Institute of Plant Nutrition, University of Hohenheim.

A cylinder of soil surrounding the root at a distance of up to 2 - 5 mm from the root surface is known as the rhizosphere (Curl and Truelove (1986)).

Conditions in the rhizosphere are very different from those in the bulk of the soil. The pH can deviate by one unit or more (Marschner et al. (1986)), microbial population density is 5 - 20 times higher than that in the bulk of the soil, and the species composition is different. The concentration of nutrients and other minerals also differs greatly from average soil values. Thus, the data from the bulk soil are not directly relevant to the root's immediate surroundings.

The roots of most, but not all, crop species are associated with a particular class of fungi: mycorrhizae. Sugar beet is a notable exception, having no such association. Mycorrhizae extend from the root surface and take up phosphorus making more of this nutrient available to the crop, in effect expanding the surface area of the roots.

This plant-mycorrhizae relationship is important for the supply of phosphorus and water to crops and forest trees, particularly where the soil's phosphorus levels are low. The relationship can be damaged by pollution, e.g., acid rain, to the detriment of the crop or tree.

However, detailed study of the plants mycorrhizal system is beset with experimental difficulties and the role of mycorrhizae in crop plants is largely unquantified.

Though increased knowledge would be welcome, present evidence indicates that correct fertilization is not detrimental to soil life but supports it (Curl and Truelove (1986, p 117), Shen et al. (1989)).

Soil microflora: bacteria, fungi

The soil's fungal biomass is similar in size to the bacterial biomass, but vary with the method of measurement. Estimates of microbial biomass range from about 800 to 3000 kg dry weight/ha in the top 20 cm of soil (plough depth). This is at least as large as the root biomass of a normal agricultural crop.

However, the activity and population of these microbes vary greatly with soil type, nutrient and carbon supply, temperature, water and oxygen supply, pH and plant community. The microflora react quickly to changes in their environment and soil bioactivity is not constant.

The microflora in the soil form an extremely complex system which presently defies detailed description and prediction. The activity can be crudely characterized by some measurements (e.g., respiratory or enzymatic, biomass and population composition); but at present, such data give little guidance relevant to crop performance and soil productivity (except for data on pest populations).

High biological activity is not necessarily beneficial, as it can be associated with excessive decomposition of soil organic matter.

Biological activity in soil is primarily determined by the supply of easily degradable organic matter. When an adequate supply is available, nitrogen may limit microbial activity for short periods. In a field situation, the major determinant of microbial soil activity is its content of organic carbon. This in turn is related to the crop rotation system used and the proportion of the crop returned to the soil.

However, some bacteria directly influence nutrient state and availability in the soil by:

- oxidizing ammonium to nitrate (nitrification)
- converting nitrate to nitrous oxide and (mainly) nitrogen gas (denitrification)
- excreting enzymes (urease) that liberate ammonia from urea
- liberating mineral nutrients from organic matter, perhaps also from inorganic minerals
- producing plant growth hormones that further root development
- competing with pathogens thereby limiting their opportunity for causing disease.

Efforts have long been made to exploit such effects, but progress is slow. An example is the use of specific chemicals (nitrification inhibitors, Section 4.3) to retard biological oxidation of ammonia to nitrate in the soil.

It is common practice to inoculate legumes with nitrogen fixing bacteria, but other cases of crop and soil inoculation with micro-organisms have given erratic results. The potential for improving yields by such means is somewhat doubtful (Campbell and Macdonald (1989)).

A factor of great influence on soil life is water supply. Excess water can induce anaerobic conditions, while drought represents another extreme type of stress on soil life. Differences in water status due to cultivation or other farm practices can be a major complicating factor in interpreting measurements of the soil's biological activity.

Reference: Paul and Clark (1989).

Microfauna

Many forms of small animals (e.g., protozoa, nematodes) in the soil feed on bacteria, fungi and roots. Through their predation on fungi and bacteria they participate in the cycling of soil nutrients. Some are plant parasites, some act as carriers for diseases. It is obvious that a low population of parasites is advantageous but most members of the microfauna have little direct influence on crops. Their numbers and activities are determined by the availability of food and are therefore mainly determined by the input of organic matter to the soil. Thus the application of manure will stimulate the growth of microfauna. Their number can be somewhat lower in fields that receive only mineral fertilizer, compared with fields which are also given manure, or which are ploughed grassland. It is however questionable if such differences in total microfauna mass are of practical significance for soil productivity.

Macrofauna: earthworms

Although earthworms are not essential for a productive soil, the presence of an active and abundant population of earthworms is generally beneficial. Their beneficial action is mainly through effects on soil structure. Earthworms feed on decaying plant residues, and on the bacteria and fungi decomposing such wastes. In this process they fragment the waste and distribute it through the soil and thus promote its further mineralization. Their burrows ease the spread of plant roots, and also facilitate air and water movement in the soil. The relative importance of these effects varies with soil characteristics, as does the size and range of species of the earthworm population.

There is much evidence to show that residues such as manure and the use of mulches (plant residues covering the soil) increase earthworm populations. Mulching can give some frost insulation and thus also increase their winter survival.

Mechanical disturbance of the soil, such as ploughing and rotavating, and compacting of soils by heavy agricultural machinery adversely affect earthworms, and can drastically reduce their numbers.

Reported effects of fertilizers on earthworm population are inconsistent.

Fertilization and cropping systems which ensure the ample return of organic residues promote earthworm populations, but acidifying fertilizers (such as types rich in ammonium) can have detrimental effects, depending on their effect on soil pH. It seems that the effect varies from soil to soil depending on its physical and chemical characteristics (Rushton (1988)). Use of pesticides may also reduce earthworm populations. Undisturbed grassland usually has a large earthworm population, provided the soil has a pH above 4.5.

Practices that promote earthworm population such as mixed animal/arable farming with extensive grasslands and manure usage, mulching, use of light tractors, and minimum tillage, are more commonly used in alternative than in intensive arable agriculture. It is therefore often observed that the latter farms have lower earthworm populations but this is not directly due to fertilizer use.

References: Hög (1985a), Lee (1985), Syers and Springett (1984).

4.3 Nitrogen

Many of the environmental issues concerning agriculture are directly or indirectly related to nitrogen.

Nitrogen occurs in various forms. Those discussed in this section are:

gases	:	nitrogen * (N_2) nitrous oxide (N_2O) ammonia (NH_3)
ions	:	nitrate (NO_3^-) ammonium (NH_4^+)
organic forms	:	urea ($CO(NH_2)_2$) natural products containing nitrogen present in living and dead organisms (e.g., proteins). nitrogen in transformed organic residues, e.g., humus.

The conversion of nitrogen between these and other related forms constitutes a complex network: the nitrogen cycle. Some background information on the nitrogen cycle is helpful in understanding the environmental issues.

The nitrogen cycle

A simplified version of this complex web of interrelated reactions is given in Figure 4.3.1. More extended versions can be found in Clark and Rosswall (1981), Söderlund and Rosswall (1982), Sprent (1988), Wilson (1988).

In order to clarify a very complex subject, we will describe the main processes in the nitrogen cycle with reference to Figure 4.3.1:

- nitrogen fixation
- conversion of nitrogen in soil
- plant and microbial uptake of soluble nitrogen compounds
- agronomic circulation and removal of nitrogen.

* *Strictly speaking, nitrogen is the name of the element. The correct scientific name for nitrogen gas in the air is «dinitrogen» but nitrogen is in common usage for this inert gas.*

Figure 4.3.1 THE AGRONOMIC NITROGEN CYCLE

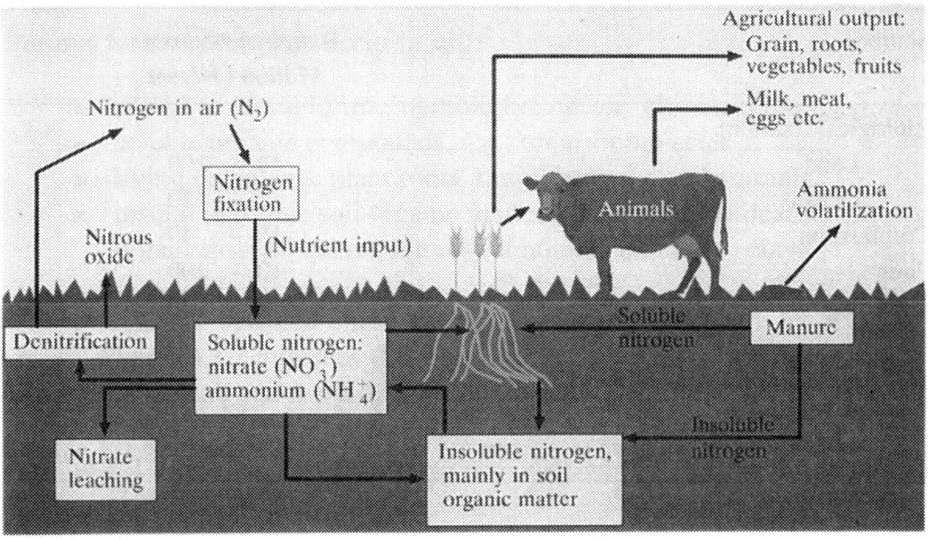

Nitrogen fixation

Nitrogen gas in the atmosphere is the ultimate nitrogen source, but it has low chemical reactivity under most natural conditions and only some bacteria can utilize nitrogen gas. To be usable by most organisms, nitrogen must be converted to ammonia or nitrate; this conversion is called nitrogen fixation. Nitrogen can be fixed:

- through biochemical processes in some bacteria (biological nitrogen fixation)
- through industrial synthesis of ammonia for conversion into fertilizers and industrial chemicals
- by heating air to a high temperature as in combustion (engines, fires) and lightning.

Various estimates exist for the amount of annual global nitrogen fixation. Industrial production of ammonia is known with reasonable accuracy; the inputs from other sources can only be estimated and estimates vary greatly. Table 4.3.1 gives the ranges of 13 different estimates listed by Söderlund and Rosswall (1982).

Table 4.3.1 ANNUAL GLOBAL NITROGEN FIXATION ESTIMATES

Source	Range of estimates Million t N/year
Biological fixation:	
Land	44 - 200
Oceans	1 - 120
Combustion	15 - 40
Fires	10 - 200
Lightning	8 - 30
Industrial ammonia production (Bumb 1989)	
Fertilizers	84
Other industrial uses	16

Industrial nitrogen fixation (Chapter 1) and combustion have substantially increased nitrogen input to the global biosphere, but the relative magnitude of the increase is somewhat uncertain. On arable land in developed nations, fertilizer nitrogen is the largest annual nitrogen input. Typical European inputs are (Section 3.2):

Fertilizer applications	100 - 250	kg	N/ha
Aerial deposition	20 - 40	"	"
Biological nitrogen fixation (soil bacteria)	5 - 10	"	"

Aerial deposits (with rain and dust) derive partly from combustion and partly from the circulation of ammonia emitted from animals and plants. Some free-living soil bacteria can fix nitrogen but the input is too small to contribute significantly to agricultural needs. Attempts to develop methods for increasing this have so far met with failure.

While grain legumes (peas and beans) remain important, the growth of legumes for their nitrogen fixing ability is not common in current farming. Fertilized grassland provides higher and more reliable herbage yields than does an unfertilized grass-clover sward. The greatest value of legumes as nitrogen sources is probably in less intensively managed grassland supporting beef and sheep production.

Conversions of nitrogen in the soil

Various forms of nitrogen occur in soil:

- soluble mineral forms: ammonium, nitrate, nitrous oxide (gas)
- soluble organic compounds, e.g., urea, amino acids
- living organisms: plant roots, fungi, bacteria, soil animals
- insoluble forms: soil organic matter (decomposable dead organisms, cell debris, humus), ammonia bonded to clays.

Transformations between these forms are mostly mediated by soil microbes. Figure 4.3.2 indicates the relationship between the various forms of nitrogen in the soil and the processes that convert one form to others.

Figure 4.3.2 CONVERSIONS OF NITROGEN FORMS IN THE SOIL

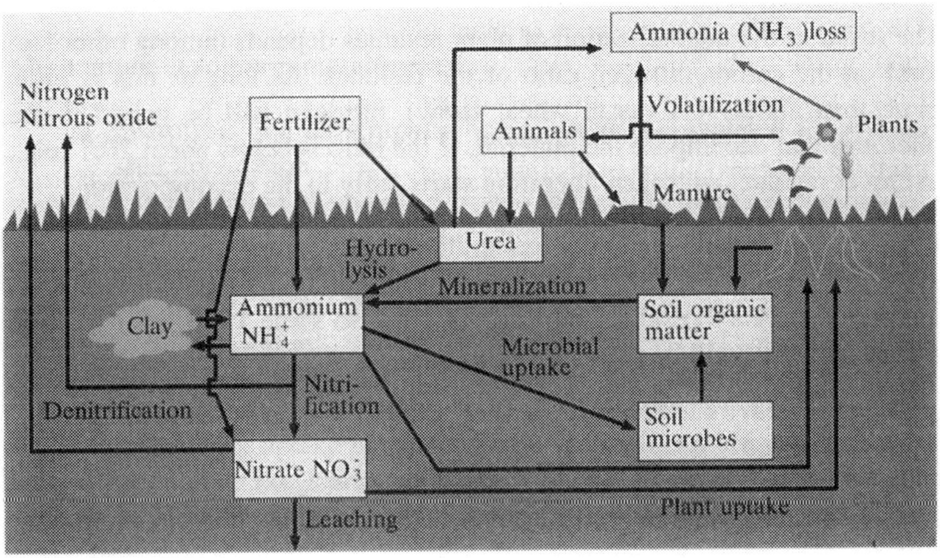

Nitrogen in soil organic matter

Soil contains large reserves of nitrogen in organic matter. In UK topsoil these are some 3000 - 7000 kg N/ha for arable land and 12 000 - 20 000 kg N/ha for old grassland. The origin of these reserves in soil organic matter (plant roots and residues, manure) is described in Section 4.2. On average, a small amount of this organic nitrogen (some 1 - 3 per cent) is mineralized each year to soluble forms through microbial action. This proportion can be greatly increased where soils rich in organic matter are cultivated. The ploughing of old grassland is a good example of this.

Under most circumstances, mineralisation is roughly balanced by the addition of organic nitrogen in roots, stubble and manures. However, small differences in amounts of mineralization and addition can accumulate over many years resulting in the gradual accumulation or loss of soil organic nitrogen. Globally, the nitrogen reserves of soil are probably declining. However, the pool of organic nitrogen in soil is so large that minor variations in its size cannot be measured with satisfactory accuracy over a few years.

Mineralization

The microbial conversion of nitrogen in organic residues and soil organic matter into soluble forms is called mineralization. In this process, carbon compounds are degraded and used as a source of energy. Nitrogen in excess of the microbial need is liberated. Soluble nitrogen forms are also liberated through the death and rupture of cells.

The result of the mineralization of plant residues depends (among other factors) on the carbon/nitrogen ratio of the residues. As long as this is high, more than about 30:1 (as in wheat straw), nitrogen will be bound in the microbes that decompose the material. If the ratio is below about 20:1 (e.g., in clover residues) nitrogen liberation starts early in the decomposition.

Plant residues can be rapidly mineralized. Depending on the crop, the amount of residues and the carbon nitrogen ratio, the nitrogen set free from the previous year's residues can range from 0 to some 150 kg per ha. The highest amounts will be present:

- after heavily fertilized vegetables when only part of the plant is harvested
- after legumes
- after ploughing of grassland, especially old land.

Mineralization of soil organic matter proceeds whenever the soil is not frozen. In Europe, the rate is often highest in the autumn when soils are relatively warm and moist and when the soil is frequently cultivated. The high soil temperatures that occur in hot climates can result in high rates of mineralization and rapid loss of soil organic matter.

Ammonium

Ammoniacal nitrogen occurs in two forms: ammonium ions (NH_4^+) and dissolved ammonia (NH_3) in equilibrium, Figure 4.3.3.

$$NH_4^+ \longleftrightarrow NH_3 + H^+$$

Figure 4.3.3 EFFECT OF pH ON RELATIVE CONCENTRATIONS OF AMMONIUM AND AMMONIA IN SOLUTION

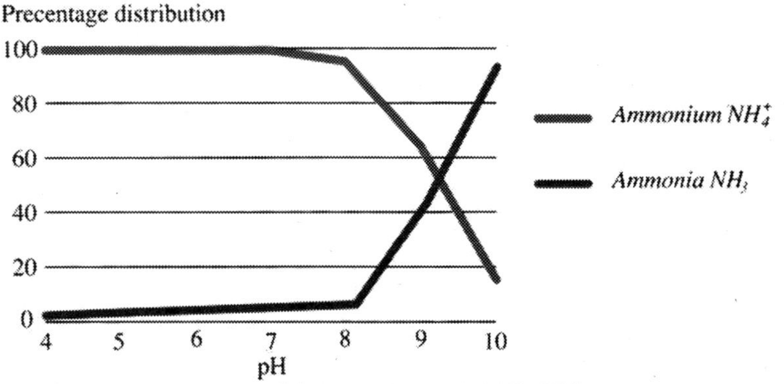

Source: Court et al. (1964).

Ammonium in soil originates from:

- hydrolysis of urea, a rapid process, which is completed a few days after application
- mineralization of organic nitrogen in manure, plant residues and humus
- application of fertilizer containing ammonium (e.g., ammonium nitrate)
- aerial deposition.

Ammonium is reversibly absorbed on soil clay particles where it remains readily available for uptake by crops. Some such mineral nitrogen is always present in the soil. For loamy soils about 20 kg nitrogen per ha typically remains in the soil at harvest time for cereals and sugar beet, crops and mown grassland.

Loss of ammonia from agriculture to the atmosphere is described below.

Ammonium in the soil is further converted through:

- uptake by plants and microbes
- conversion to nitrate (nitrification).

Nitrification

Ammonium is oxidized to nitrate by soil bacteria of the genera *Nitrosomonas* and *Nitrobacter*. This process is termed nitrification. Nitrite is formed as an intermediate; it is rapidly further oxidized to nitrate. Nitrous oxide is also formed. Nitrification is normally a fairly rapid reaction, being completed in one to a few weeks. The reaction rate is greatly retarded in acidic soils (pH below 5.5) and at low temperature (below 4 - 5°C).

As nitrification is principally performed by a very specialized set of bacteria, it is possible to delay the process a few weeks to a few months by specific chemical nitrification inhibitors (e.g., nitrapyrin, dicyandiamide) acting on these organisms. Such inhibitors can be effective when incorporated into manure or fertilizers containing ammonium. But they cannot be used for delaying nitrification of ammonium derived from mineralization of soil organic matter due to the large amounts that would be needed. Reported results from the use of nitrification inhibitors are not always consistent. It is presently unclear if they should be regarded as fertilizer additives or as agrochemicals.

Nitrate

Nitrate is soluble in water and only weakly bonded by soil particles. It is thus a very mobile form of nitrogen in the soil, moving with the flow of water and by diffusion. The main sources of nitrate are nitrification, fertilizer application and aerial deposition. Nitrate is a principal nutrient for plants and microorganisms. It can also be lost from the soil by leaching and by denitrification.

Denitrification

Denitrification is the reduction of nitrate to nitrous oxide and nitrogen gas. It is the ultimate process for the return of fixed nitrogen to the atmospheric pool.

Some bacteria can use nitrate as a source of oxygen when conditions become oxygen deficient (anaerobic). Denitrification therefore takes place mostly in waterlogged situations, though anaerobic conditions can exist as small pockets in well aerated soils. Denitrification requires readily degradable organic matter as an energy source and its application can enhance the process.

Denitrification also occurs through chemical reactions with iron compounds in deep soil layers.

The formation of nitrous oxide and related issues are discussed in Section 2.3.

Reference: Paul and Clark (1989).

Plant uptake of soluble nitrogen compounds

Plants take up nitrogen compounds both as nitrate (usually the dominant form of soluble nitrogen in the soil) and as ammonium. The balance varies with species and circumstances, but in general nitrate is the major source of plant nitrogen. There is some evidence that very small amounts of urea and amino acids can be taken up by plants from the soil solution.

Bacteria and fungi can also use both ammonium and nitrate as nitrogen sources but generally prefer ammonium.

Plant uptake of nitrogen can be studied through the use of fertilizers enriched in the stable nitrogen isotope N^{15}. The results vary within wide limits. Typical results are illustrated by French reports on the fate of the nitrogen given to a crop, Table 4.3.2.

Table 4.3.2 FATE OF FERTILIZER NITROGEN GIVEN TO A CROP

	Per cent
Taken up by the crop (above ground parts)	40 - 60
Incorporated in the soil's organic matter	20 - 50
Mineral form in soil (clay-ammonium complex)	5 - 20
Lost by denitrification and volatilization	2 - 30
Lost by leaching	2 - 10

Source: Machet (1987), Machet et al. (1987).

The proportion of applied nitrogen taken up by the crop is affected by many factors including crop species, climate and soil conditions. Recovery in the crop of up to 86 per cent of the nitrogen applied to winter cereals has been reported (Jenkinson (1982)).

The rate of nitrogen uptake varies with the stage of plant development. Wheat takes up most of its nitrogen needs in the first half of the year while potatoes have a more even uptake through the season (Section 2.3).

A plant's capacity for depleting the soil of nitrate depends on many factors:

- length of growing season (grass is efficient in nitrogen uptake due to its long growing season)

- rooting depth and root density (grass has a high root density to intercept nitrogen added to the surface while winter cereals, oilseed rape and sugar beet have deep roots to exploit the lower soil layers)
- availability of other nutrients (deficiency of phosphorus for example will restrict nitrogen uptake)
- incidence of disease (fungal diseases of crops can reduce nitrogen uptake)
- soil moisture (excessively dry or wet conditions will restrict nitrogen uptake).

Circulation and fate of nitrogen in agriculture

There is a substantial circulation of nitrogen in agriculture: harvested crops are fed to animals and the manure is returned to the soil.

In 1978 about 80 per cent of all nitrogen in UK crops was given to animals for feed; only 20 per cent was consumed directly by humans. These proportions will be similar today.

Sow with hybrid piglets from the crossing of different breeds.
Photo: N. Standal, Agricultural University of Norway.

Of the nitrogen in animal feed only 18 per cent is passed on as milk and dairy products, meat, wool, etc. The main part, 82 per cent, is in animal waste (Royal Society (1983)), mostly excreta. Dutch reports are similar: nitrogen output with milk from a dairy farm was 15 per cent of the input. The animal husbandry sector and its production and use of manure are thus a major factor in the nitrogen cycle in agriculture, and also a major source of nitrogen losses.

An estimate of the nitrogen balance of UK agriculture is presented in Table 4.3.3.

Table 4.3.3 NITROGEN BALANCE OF UK AGRICULTURE

Estimated inputs and outputs of nitrogen in agricultural land of the U.K. in 1978:

Input	1000t	Output	1000t
Rain	275	Crops and grass	
Seeds	14	(mostly for animal feed)	1367
Fertilizers	1150	Leaching	326
Sewage	26	Ammonia volatilization	
Livestock excreta	1020	from livestock excreta	536
Silage effluent	9	crop wastes	50
Straw	15	sewage	9
Feed waste	9	Balance by difference	
Biological N_2 fixation	150	(mainly denitrification	
		and immobilization)	380
Totals	2668		2668

Source: Royal Society (1983).

A similar balance for Norwegian agriculture is given in Figure 4.3.4. The figure illustrates the complexities of the nitrogen cycle in agriculture.

Figure 4.3.4 NITROGEN BALANCE IN NORWEGIAN AGRICULTURE

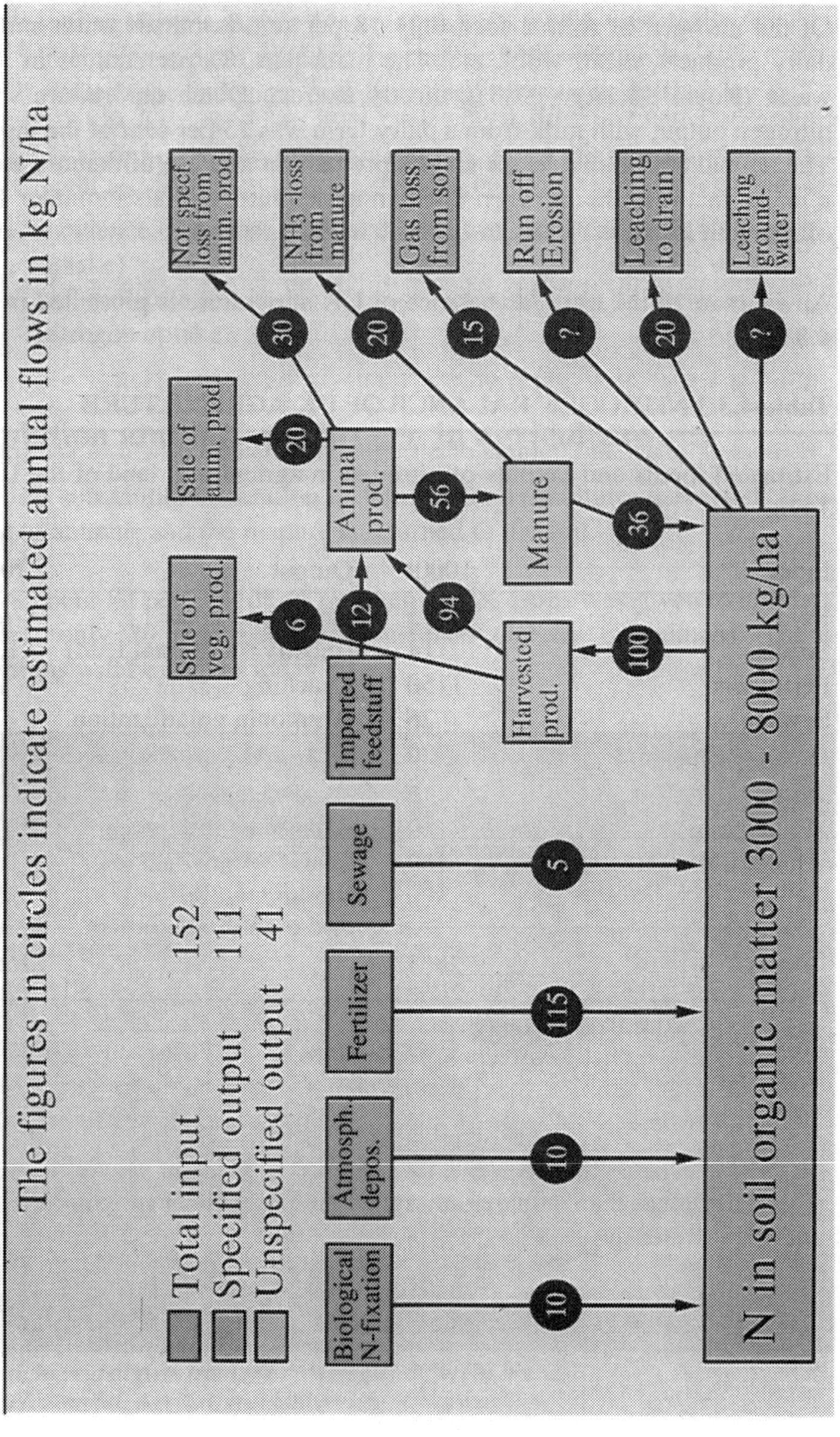

Principal issues relating to nitrogen use in agriculture

There are many issues relating to nitrogen use in agriculture. Topics such as formation of nitrous oxide, effects on plant susceptibility to disease, crop quality, resource conservation and eutrophication are treated in other sections.

In this part we will discuss:

- the low overall efficiency of agricultural use of nitrogen
- losses: - ammonia to the atmosphere
  - nitrate in water and related issues.

The low overall efficiency of nitrogen input

A current issue is the fate of nitrogen input. British, German and Danish estimates indicate that only about 10 - 30 per cent ends up in food.

There are three main origins of nitrogen losses in agriculture:

- animal husbandry
- field operations
- food transport, storage and processing.

As previously indicated, the animal husbandry sector is the main source of this apparent inefficiency in nitrogen use. The losses originate mainly from:

- ammonia volatilization from animals, their excreta and from manure storage and application
- inefficiency of manure as a nitrogen source through its mineralization outside the main growing season and consequent losses
- regional imbalances between manure production and needs, giving waste and excessive applications (e.g., in Southern Netherlands, Brittany in France, parts of Jutland in Denmark).

Much of the fertilizer used is applied to crops destined to be used as animal feed. Thus a major part of the nitrogen in manure originates as fertilizer nitrogen but losses in the animal sector will not be reduced by attention to fertilizer use on feed crops. Such losses must be controlled through developments in the animal sector itself.

Increasing nitrogen efficiency in the animal husbandry sector is a major challenge in the development of an agriculture with increased sustainability. Even if progress is made, nitrogen losses are likely to remain associated with animal manure.

Nitrogen losses in other field operations in agriculture are:

- denitrification
- volatilization of ammonia from crops
- nitrate losses through leaching.

The first topic has been discussed previously in this section. The two latter are treated below.

Even if efforts are made in all sectors to increase the efficiency of nitrogen use in agriculture, the complexity of the nitrogen cycle implies that substantial overall nitrogen losses in agriculture are unavoidable.

Ammonia volatilization and deposition in agriculture

Soils and crops obtain part of their nitrogen supply from ammonium and nitrate in rain water and through dry deposition of nitrogen compounds. In the long-term experiments at Rothamsted for the period 1852-1967, the steady-state annual removal of nitrogen with the crop from non-fertilized plots was 30 kg N/ha.

Nitrogen inputs by this route have increased since the Second World War, e.g., by some 5 kg N/ha over much of Southern and Eastern England. The increase and its composition vary regionally. It is a cause of concern because of:

- eutrophication of seas and fresh water (Section 4.8)
- soil acidification as bacteria in the soil oxidize ammonia to nitric acid (Section 4.2)
- increased nutrient supply to ecosystems naturally poor in nutrients, such as heathland. This can cause the decline and extinction of plant species adapted to low nitrogen supply
- increased nitrogen supply to forests possibly contributing to forest death, e.g., through delaying the onset of autumnal processes leading to winter hardiness
- increased nitrate leaching from some forests.

The increased nitrogen deposition originates from:

- nitrogen oxides from high temperature combustion, e.g., in cars, electricity generation: some 3.6 million tonnes nitrogen annually in Western Europe
- ammonia emissions: some 2.9 million tonnes nitrogen annually (Western Europe) (Buijsman et al. (1986), Eliassen et al. (1988)).

The ammonia emissions come mainly from animal husbandry. Currently some 40 per cent of the nitrogen taken in by livestock is lost through ammonia volatilization from the animals themselves and, mainly, from losses during the collection, storage and application of manures (Table 4.3.3). Such ammonia losses are estimated at around 2.5 million tonnes N per annum in Western Europe. Not all of this is completely lost as part at least will eventually return to agricultural land by aerial deposition.

Recently, maturing cereal crops have been recognised as a significant source of ammonia emission. Most of the nitrogen in grain is translocated from other parts of the cereal plant, a process which is associated with ammonia losses. Factors influencing this process are incompletely known and research is continuing (Wetselaar and Farguhar (1980)).

Under acidic conditions, ammonia is bound as ammonium. The loss of ammonia from soil at normal soil acidity is therefore insignificant. But soil injection of ammonia and surface application of urea (especially on calcareous soils) can result in loss of ammonia from volatilization. On the European scale, these two processes are only minor sources of ammonia emissions to air. Thus efforts towards the reduction of these emissions are unlikely to contribute significantly towards a reduction in present levels of ammonia deposition from the air.

Nitrate leaching

In many countries, the concentration of nitrate in drinking water is covered by regulations. Within the EC, the limit is 50 mg nitrate per litre. Nitrate concentration is at or approaching this limit in some parts of Europe and the USA. For example, in France around 1 million and in the UK around 1.6 million people are presently supplied with water which can at times exceed the statutory limit. There is also concern about nitrate input to coastal waters from rivers as a possible contributor to eutrophication (Section 4.8).

Nitrate in ground water in Europe comes mainly from agricultural operations. Nitrate in rivers and surface waters originates from agriculture and from urban sewage.

Leaching of nitrate is a principal environmental and health issue related to fertilizer use. Current disagreements centre not on whether measures should be taken to prevent deterioration of water resources, but on what measures are appropriate when effects and benefits are taken into consideration.

Nitrate and water movement in soil

Nitrate in soil follows the movement of water but will also migrate by diffusion. Plant roots are active in nitrate uptake and deplete the root zone of nitrate. This is replenished by diffusion from the surrounding soil solution, even though water movement may be in a different direction.

Sampling of soil water at various depths provides information about nitrate leaching. Photo: Hydro Agrar, Hanninghof.

Water movement in soil depends on soil structure and porosity, water supply from precipitation and irrigation, evaporation from the soil surface, and the degree of drainage. Plant uptake of water exerts a major influence. Cracks and wormholes can permit some rapid percolation of surface water to deeper layers even when the general soil water movement is upward due to evapotranspiration.

During the growing season more water is used by the plant than is precipitated. There will then be an upward movement of water in the soil which brings with it nutrients, especially nitrate. In this manner, nitrogen that has temporarily been out of reach for plant uptake can again be made available. The amount of upward water transport depends upon soil type and is greatest in loamy soil.

In late autumn, winter (except where the ground is frozen) and early spring, the principal direction of water movement is towards ground water reservoirs and drainage systems. Then leaching can occur.

Water movement in the deeper parts of many soils is usually slow, from some centimetres to a few metres per year, depending upon soil type. The content of nitrate at greater depths can therefore reflect the events of years ago. All other factors being equal, higher rainfall will tend to be associated with greater leaching of nitrate. In dry areas, nitrate can accumulate in the soil and be available for the following season's crop.

Land management and cropping systems

Land that is fallow for part of the year has a greater potential for leaching than land under plant cover most of the time. This is exemplified by data from Switzerland (Table 4.3.4); there are similar reports from other countries.

Table 4.3.4 THE INFLUENCE OF CROP COVER ON NITRATE LEACHING

Culture	Fallow	Maize	M+R	Grass
Fallow part	100 %	Large	Small	0 %
N-fertilizer, kg N/ha	0	120	120	250
N-leaching, kg N/ha				
Clay	100	72	27	8
Sand	167	60	24	9

Rotation: M + R = maize + rape (autumn sown). Source: Furrer (1984).

Such experience forms the basis for the principal measure to lessen nitrate leaching: keeping fields green with winter or catch crops as far as practicable. Such crops reduce leaching if they are given sufficient time to develop their root system before winter slows down growth. Intercropping (planting a main crop, e.g., cereals together with a catch crop, e.g., grass) is a current field of development. In intercropping a grass sward develops after grain harvest. Such combinations are interesting but carry a yield penalty and can be complex and difficult to manage.

Grasslands with a very long season of nitrate uptake and a dense and extensive root system are usually not prone to leaching. But on grazed grassland, dung and urine will give spots rich in mineral nitrogen. This can result in nitrate leaching even when the grass has been given less than optimal amounts of fertilizer. Hence in grazed grassland there is a higher risk of leaching than from mown grassland.

Old, permanent grassland releases vast amounts of nitrate when ploughed. One such site at Rothamsted lost nearly 4000 kg nitrogen per ha, mostly as nitrate, during the first 18 years after it was ploughed. Samples from the chalk beneath the field indicate that for a number of years the water draining from the soil had a nitrate content three times higher than current EC limits.

Some field vegetables need a high content of mineral nitrogen in the soil solution in order to yield satisfactorily. For example, cauliflower contains only a small amount of nitrogen in the produce and most of the fertilizer applied remains in the soil and in plant residues. In such systems, leaching can be substantial. It is also difficult to prevent extensive leaching from vineyards on steep slopes and with bare soils between the rows.

Leaching can be reduced by ploughing in straw and crop residues which contain little nitrogen. Their decomposition temporarily immobilizes mineral nitrogen as microbial biomass. On the other hand, crop residues such as those of oilseed rape that are rich in nitrogen can contribute to nitrate leaching when ploughed in.

Ploughing and related methods of cultivation increase the mineralization rate and the risk for nitrate leaching. Frequently ploughed soils are vulnerable to erosion which also leads to loss of nutrients.

Nitrate leaching with different nitrogen supply sources

Mineral fertilizer

Application of nitrogen fertilizer increases the concentration of ammonium and nitrate in the plough layer of soil. Springtime rain can leach nitrate out of reach of the roots. This risk is greatly reduced as the crop roots develop in mass and depth. It can be advisable to give the nitrogen in split applications during the growing season.

Crops such as cereals, grass and sugar beet have deep and extensive root systems that effectively deplete the soil of soluble nitrogen. Fertilizer recommendations are set to achieve economic optimal yields. Measurements show that such fertilizer rates leave little or no soluble nitrogen in the soil at the end of the growing season for such crops. This is exemplified by Danish results, Figure 4.3.5. Similar results are reported by others (Dilz, (1988)).

Figure 4.3.5 THE INFLUENCE OF FERTILIZATION ON YIELD AND LEACHING

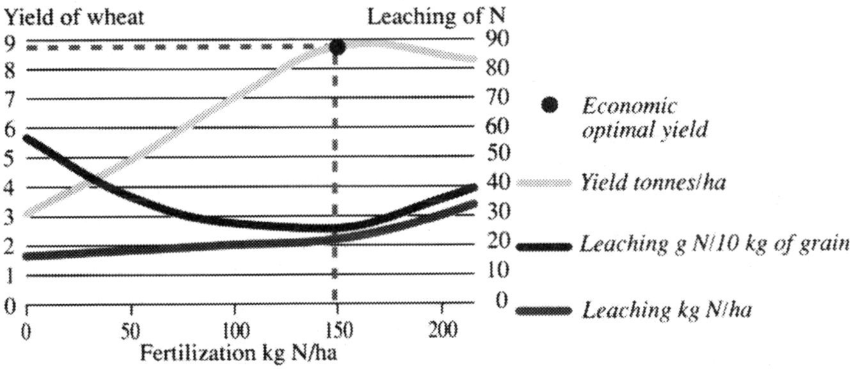

Based on data from Danish Goverment National Committee on Crop Husbandry (1985).

It was found that the amount of nitrate leached rises slowly compared with the increase in yield up to about the fertilizer rate which gave the optimal yield. This is also the rate that gives the minimum amount of nitrate leached per produced crop unit. But at excessive rates of fertilizer application, leaching can increase markedly.

In contrast to cereals and sugar beet, residual mineral nitrogen after harvesting field vegetables and potatoes is substantially higher after optimal fertilization than at sub-optimal rates (Kuhlmann and Engels (1989)). This is mainly due to the low rooting density and shallow root systems for these crops. High amounts of residual nitrate are especially found when such crops are harvested during their rapid phase of growth, when they require high nitrate concentrations in the soil to attain high yields.

With this exception, almost all nitrate leached over winter comes from the mineralization of soil organic matter. Where fertilizer nitrogen is applied at rates that do not exceed the economic optimum, its direct contribution to nitrate leaching is small (Macdonald et al. (1989)).

Intensive agriculture with adequate fertilization can leave increased amounts of nitrogen in soil organic residues, compared with farming in former times when nitrogen was in short supply. This will enhance soil fertility but also increase the potential for nitrate leaching from mineralization.

Mineralization is also the origin of leached nitrate from fields that obtain their nitrogen from legumes or manure.

In these circumstances, management practices to minimize nitrate leakage from arable land should achieve best results by minimizing the soil's nitrate accumulation from mineralization after the autumn harvest.

Manure application

In manure, part of the nitrogen is present in a water soluble form, mainly urea and ammonium. This will act in the same manner as fertilizer nitrogen. The organic nitrogen becomes available only after gradual mineralization. This will extend over many growing seasons. Application of nitrogen

Figure 4.3.6 UTILIZATION OF MANURE COMPARED TO FERTILIZERS IN CROP PRODUCTION

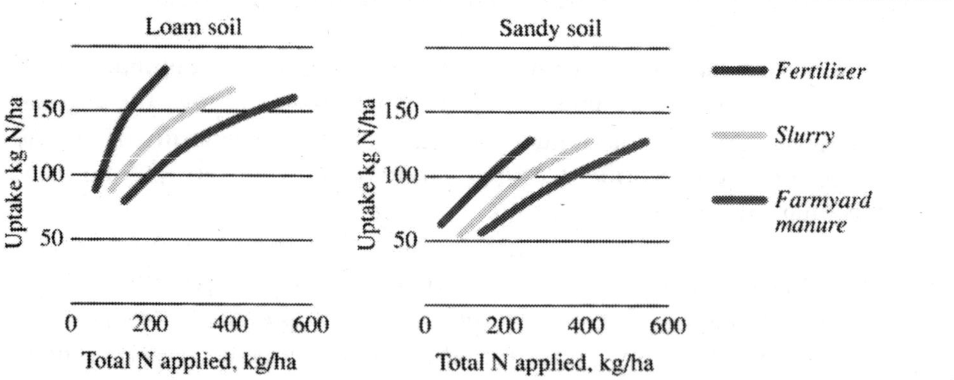

Source: Kofoed (1985).

as manure must therefore be greater than that of nitrogen as fertilizer in order to give the same yields, Figure 4.3.6. Some of the residual organic nitrogen in manure will be mineralised in autumn and at risk of leaching over winter.

Nitrate leaching originating from manure application is a problem in regions with high stocking rates in relation to land available for spreading manure.

Legumes as nitrogen source

Legume residues can provide nitrogen, but only after mineralization.

Danish experiments exemplify the potential for nitrate leaching when clover-grass swards are ploughed (Table 4.3.5, column for no nitrogen application).

Table 4.3.5 NITROGEN LEACHING, in kg N/ha • year*

The effect of breaking clover-grass swards, and increasing fertilization rates.

Year	Crop	Fertilization No N	1/2N	1 N**	3/2 N	mm Water drained
78/79	Barley/Clover grass undersown	11.7	9.1	9.8	14.3	67
79/80	Clover-grass	10.9	7.9	9.5	19.8	205
80/81	Clover-grass	18.8	13.2	24.4	77.0	425
81/82	Barley	41.8	43.9	44.8	48.5	272
Average 4 years, kg N/ha • year		20.8	18.5	22.1	39.9	242

* Measurement period 01.04 - 31.03.

** 1N = Economic optimal fertilization rate.

The fields were ploughed in the autumn of 1981 and spring barley was sown in 1982.

Source: Kjellerup (1983).

Italian ryegrass under-sown in wheat. When the wheat is harvested, a grass sward develops. Such a catch crop can help control nitrogen leaching. It is ploughed down next spring as green manure for next crop.
Photo: T. A. Breland, Department of Biotechnological Sciences, Agricultural University of Norway.

Agricultural practices that minimize nitrate leaching

It follows from the preceding discussion that nitrate leaching depends on many factors besides fertilizer rate, e.g., soils, crops, rotations and weather patterns. Generally application of the following principles of good agricultural practice, suitably adapted to local conditions, should reduce nitrate leaching to a practical minimum:

- fallow periods should be avoided. The soil should be kept under green cover for as much of the year as possible, by early sowing of winter crops or intercropping
- legumes should not be ploughed down before winter. They should preferably be grown so that a subsequent crop immediately following can take up the nitrogen released by mineralization of the residues
- grassland should not be ploughed until shortly before the next crop can be established, and old grassland should preferably not be ploughed. Where such ploughing is necessary, rapid establishment of plants with a high requirement for nitrogen is especially important
- soil tillage should be minimized and preferably avoided in the *autumn*
- straw should not be burned or removed but ploughed in or used as mulch
- slopes should be cultivated in such a way as to minimize surface run-off
- manure should be evenly spread. Application in autumn or winter should be avoided
- fertilizer nitrogen should not be applied in autumn
- fertilizer and manure should be applied at times and in amounts appropriate to the nutrient requirements of the crop, taking into account the amount of available nitrogen already present in the soil.

While these practices are designed to minimize nitrate leaching, it should be kept in mind that some of the measures may enhance other problems such as disease incidence and weed infestation, or cause cultivation difficulties such as delayed drying out of the soil in springtime.

General references to nitrate leaching:
ECETOC (1988), House of Lords (1989), Jürgens-Gschwind (1989), Owen and Jürgens-Gschwind (1986), Prins et al. (1988), Strebel et al. (1989).

Nitrate toxicology

The issue of nitrate leaching into drinking water sources is closely connected with the subject of nitrate toxicology.

The nitrate ion (NO_3^-) as such is hardly toxic at all: on intake it is mostly and rapidly excreted in urine without interfering with body or cell processes. The concern is with nitrite (NO_2^-) which is formed by bacterial reduction of nitrate in the body: in the mouth, stomach and lower part of the intestine.

We all take in nitrate with our food and drink. Nitrogen dioxide in the air will, on absorption through the lungs, give nitrate, but the quantity is insignificant compared with that in food and water.

Nitrate is a normal cell constituent in all crops. Some vegetables (red beets, spinach, lettuce, etc.), may contain above 1 g NO_3^-/kg. Others (beans, peas, potatoes) mostly have low nitrate content, as do cereals. Vegetables are the major source of dietary nitrate intake, but the amounts vary greatly from day to day depending on diet. On average, adults living in the Western industrialized nations have an intake of about 70 mg NO_3^-/day from vegetables; vegetarians take in about 3 times as ..ch.

Drinking water can represent another important source. When the nitrate content of water is 50 mg NO_3^-/l (the EC and WHO limits) nitrate intake from this source also reaches about 70 mg NO_3^- /day.

Finally, the body produces about 30-60 mg NO_3^-/day as part of normal metabolism. This synthesis can be greatly enhanced during bowel infections but detailed studies on this are currently lacking.

Nitrate is absorbed into the blood stream from the small intestine, and most of it excreted rapidly within 24 - 48 hrs in the urine. There is no build up of nitrate in the body.

About 20 per cent of nitrate intake is excreted in the saliva and partly reduced to nitrite in the mouth. The amount produced is currently estimated at about 5 mg NO_2^-/day. Nitrate is also excreted into the stomach as a constituent of gastric juice. Nitrate can be reduced to nitrite in the stomach by bacteria, but the relative contributions of the several sources to nitrite in the stomach and the factors controlling these contributions are at present incompletely known. It appears that little nitrate normally reaches the large intestine, where reduction to nitrite and denitrification can also take place.

Concern about nitrate intake is due to:

- concern for infant health, as infants' feed made with water with more than 50 mg NO_3^-/l is believed to involve a risk of attacks of acute infant methaemoglobinaemia ("blue babies")

- concern that nitrite may react with food components in the stomach to give carcinogenic compounds such as nitrosamines, thus causing cancer, especially of the stomach but also of the liver and esophagus

- concern that nitrate may cause a variety of other diseases: goitre, serious malformations and heart disease.

Infant methaemoglobinaemia

Nitrite is one of the substances that transforms oxyhaemoglobin (the oxygen carrier in the blood) into the inactive form methaemoglobin. This transformation is a normal body process; there are enzymes present in the blood cells that reconvert methaemoglobin to haemoglobin. A normal methaemoglobin level (part of the haemoglobin pool in methaemoglobin form) is about 0.5-2 per cent. When the level exceeds 10 per cent, clinical symptoms (such as grey or blue skin) appear and the condition is called methaemoglobinaemia. With methaemoglobin levels of more than about 40 per cent the condition is dangerous, though recovery is known to have occurred in cases with levels about 70 per cent. No disabilities are noted subsequent to recovery from acute attacks.

The condition "Acute infant well-water methaemoglobinaemia" has been associated with the use of nitrate-rich well waters (usually with more, often much more, than 100 mg NO_3^-/l) for making baby feed. These well waters were also often (though not always) polluted with bacteria, or they offended against other hygienic criteria. About 3000 cases have been noted worldwide, mostly in the period 1950 - 70 .

The disease has almost completely disappeared from Western Europe and the USA, but cases seem still to occur in Eastern Europe, especially in Hungary.

Bottle fed babies get all their nitrate intake from water as milk contains practically no nitrate. It is believed that the disease is caused by bacterial reduction of the water nitrate to nitrite in the alimentary system.

It has for many years been accepted that water with more than 100 mg NO_3^-/l should not be used as drinking water or for baby feed preparation. The current European and WHO water quality regulations and recommendations are more stringent, with 50 mg NO_3^-/l as the maximum, and, for EC, 25 mg NO_3^-/l as "guide level". These limits are principally set to protect against methaemoglobinaemia.

It is, however, now known that acute methaemoglobinaemia is a rare but well documented complication in cases of infant diarrhoea. It appears now that many, perhaps all, of the methaemoglobinaemia cases ascribed to the use of water with less than 100 mg NO_3^-/l in reality could have been ascribed to other causes, principally infant enteritis (Bøckman and Bryson (1989)).The elevated methaemoglobin levels claimed for groups of infants given milk preparations made with water with 50 - 100 mg NO_3^-/l are still within normal physiological limits.

Consequently the fear that the use of water with 50-100 mg NO_3^-/l implies a real risk of infant methaemoglobinaemia seems exaggerated. The current limit of a maximum of 50 mg NO_3^-/l should represent a considerable margin of safety.

Two other sets of circumstances that have caused methaemoglobinaemia should be mentioned. Some vegetables, e.g., spinach and carrots, can have high nitrate content (Section 5.5). Bacterial spoilage of infant food made with such vegetables can reduce the nitrate to nitrite. Such spoiled food has given acute methaemoglobinaemia. Improved hygiene and control of nitrate content in the production of infant food has almost completely eliminated this problem, but very occasionally cases occur due to home made preparations.

Methaemoglobinaemia can also occur in animals, from grazing herbage rich in nitrate. The strongly reducing conditions in the rumen of cattle and sheep convert ingested nitrate to nitrite. Thus ruminants are more susceptible to nitrate poisoning than monogastric animals. Excessive application of manure or fertilizer can result in feed with potentially hazardous nitrate content. This hazard can be avoided by following recommended practices for fertilizer use on feed crops and herbage.

Nitrate and cancer

Nitrite is chemically reactive and can react in the stomach, e.g., with food components to give carcinogenic compounds. Nitrite may also react with "scavengers" to give harmless compounds, and it seems that these latter reactions dominate. Best known among the scavengers is vitamin C (ascorbic acid).

The scientific hypothesis that carcinogenic nitrous compounds can be formed in the body and may play a role in carcinogenesis has been very fruitful and has generated a large amount of work, but some of the results and conclusions are conflicting. It is at present far from clear if carcinogenic nitrous compounds are formed in the stomach in meaningful quantities when compared with the intake of carcinogens in normal food and possibly with carcinogenic metabolites from the large intestine. It is also not evident that differences in nitrate intake from nitrate contents in water and food have a meaningful influence, or whether the important factors are infections and stomach bacterial populations that can make such compounds.

Unstable carcinogens formed in the stomach should mostly affect this organ. However, the incidence of stomach cancer in the industrialized world has steadily decreased for more than a generation while nitrate exposure (through more use of vegetables, and in some areas also through water) has probably increased. The decrease in incidence of stomach cancer is ascribed to more varied diets, rich in fruit and vegetables, with less use of salt as a preservative, and to similar factors. But it remains a possibility that nitroso compounds formed in the body can be one of many factors involved in the genesis of some other cancers.

Differences in nitrate intake do not result in apparent differences in cancer risks. There are many epidemiological studies published on this subject, but results are conflicting and some of the studies are open to criticism on methodological grounds. However, recent epidemiological studies provide no evidence that dietary nitrate induces cancer in man.

In 1985 the WHO concluded: "No convincing evidence of a relationship between gastric cancer and consumption of drinking-water containing nitrate levels up to 10 mg N/l (45 mg NO_3^-/l) has emerged. Furthermore, no firm epidemiological evidence has been found linking gastric cancer and drinking-water containing higher levels of nitrate, but a link cannot be ruled out due to the inadequacy of the data available. Gastric cancer is declining in most countries, and the risk from nitrate, if any, would appear restricted to individuals with conditions associated with low gastric acidity, rather than to the population in general. Very few studies have considered human cancers other than gastric in relation to nitrates, and none of them provides convincing evidence that nitrate ingestion influences cancer incidence at other sites" (WHO (1985)).

This still summarizes the present complex situation with the theoretical possibility of, but so far no hard evidence for nitrate/nitrite involvement in cancer genesis (Forman (1989)).

It is noticeable that workers in nitrogen fertilizer plants have increased nitrate intake through dust exposure but do not show elevated incidences of cancer compared with other worker populations. Hence, sweeping statements such as "nitrate gives you cancer" are not supported by experience.

Other diseases

There are unconfirmed claims that nitrate intake is associated with other diseases:

- goitre (in East Germany)
  Nitrate in animal experiments can interfere with iodine uptake, but no verified epidemiological evidence is available
- serious malformations
  Australian claims that nitrate in drinking water causes malformations have been criticised and retracted, but the theme is still being discussed
- heart disease
  Many factors, including nitrates, have been proposed to account for cardiovascular diseases, but by far the most important are smoking and dietary fat intake.

References: ECETOC (1988), van Duijvenbooden and Matthijsen (1989), WHO (1985b).

4.4 Phosphorus*

Phosphate in soil

Phosphorus in the chemical form of phosphate is one of the principal nutrients for plants. The first fertilizer produced was superphosphate (1843). Phosphorus deficiencies were then common in Europe and elsewhere. Long-term fertilization has improved the phosphorus status of much of Europe's arable land to the extent that over large areas only maintenance application is now required. But the early development of plants often benefits from some phosphorus application even where the phosphate status of the soil is high. If fertilization with phosphorus is neglected, deficiencies gradually develop. In other parts of the world, phosphate deficiencies are not uncommon.

Topsoils typically contain some 1.3 to 3 tonnes phosphorus per ha, but only a fraction will be potentially available to plants each season.

Phosphate is supplied to cropped land at application rates ranging from a few kg phosphorus per ha to 35 kg/ha or sometimes more. Fertilizers are the usual source but manure also contains phosphorus. The use of ground rock phosphate as a nutrient source is discussed in Section 5.3.

Figure. 4.4.1 PHOSPHORUS IN SOIL

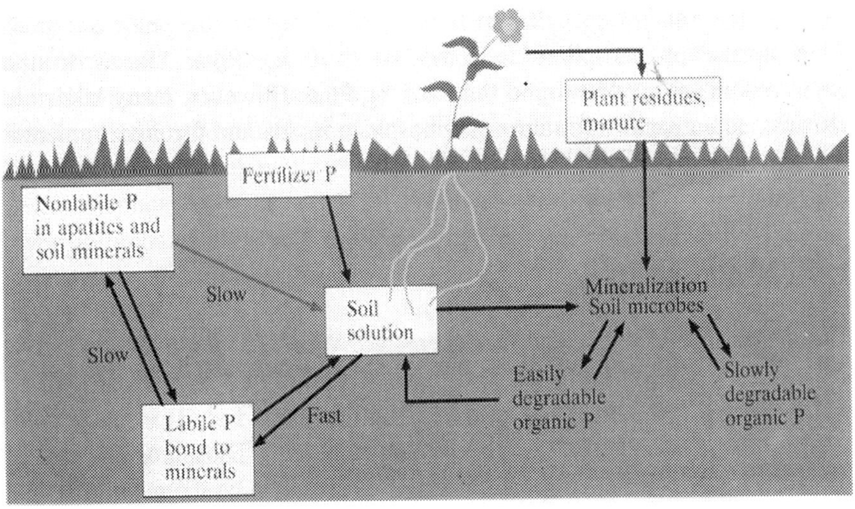

* *Phosphorus is expressed as the element P. For conversion to P_2O_5, multiply by 2.3.*

Apart from the phosphorus in living organisms, phosphorus in soil is present as:

- phosphate dissolved in soil water
- labile phosphate absorbed to soil particles, especially clay
- non-labile phosphate in soil minerals and inorganic precipitates
- organic phosphate compounds.

The relationship between these forms is illustrated in Figure 4.4.1.

The change from one form to another is reversible but the rates of conversion and reconversion vary widely. The phosphate concentration in soil water is low (in the order of 0.03 - 0.2 mg/l). It is rapidly depleted by growing plants and replenished from the pool of labile phosphorus within a time span of hours to days.

Non-labile phosphate in soil minerals and organic matter is liberated at slow rates. It takes weeks or months for such phosphate to become available and some may be unreactive within practical time limits. These rates depend on a variety of soil factors and soil acidity is especially important.

The release of non-labile phosphate to soil solution is too slow to furnish the plants with adequate amounts of phosphate. The annual application of soluble phosphate is therefore needed on most soils to maintain an adequate level of labile phosphate.

Crop uptake of phosphate is some 10 - 30 kg P/ha. Losses to the environment are small, around 0.2 - 0.5 kg P/ha. However, many lakes are sensitive to eutrophication through phosphate inputs and for environmental reasons these losses can be important.

Loss of phosphate

The main cause of phosphate loss is erosion. When soil particles, e.g., clay, are eroded from soil surfaces, phosphate is transported on the particles to water courses. Because the topsoil (plough layer) is richest in phosphate, losses can be substantial where erosion is severe. Thus erosion control measures such as grass strips along water courses can be the most effective ways of preventing phosphate loss.

While erosion of soil material is the main route for the loss of phosphate, leaching can be important in some soils.

Leaching of inorganic phosphate through soils is negligible on most mineral soils. In Figure 4.4.2 the migration of phosphate down the soil profile is shown for plots given 33 kg P/ha each year for 100 years, compared with unfertilized land. The enrichment was hardly noticeable below the plough layer. But there was somewhat more migration under grass than under cereals.

Figure 4.4.2 SOIL PHOSPHORUS PROFILE AFTER 100 YEARS FERTILIZATION

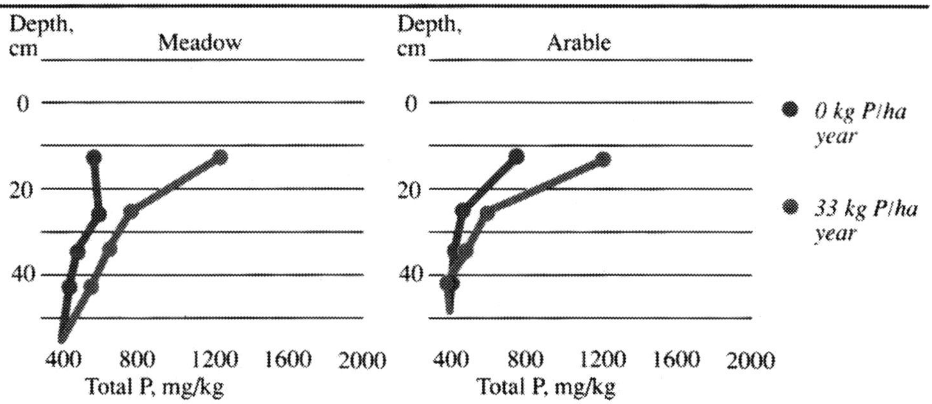

Source: Cooke and Williams (1970).

There are some exceptions to the general rule that phosphate is immobile in mineral soils. Very light sandy soils will not bind as much phosphorus as normal mineral soils. When large amounts of phosphate fertilizers have been applied to these soils just before heavy rainfall, mass movement of dissolved phosphate may occur before interaction with the soil takes place. However, phosphate leached from the upper horizon may be retained in lower horizons of the soil if these are enriched with clay. Dissolved phosphate can be transported through cracks and wormholes into the drainage system. Phosphate can also leach through organic soils which have only limited binding capacity for phosphate.

In some cases such as on grasslands and in no-till systems, phosphate is spread on the surface. Such phosphate can be at greater risk of loss than that incorporated in the soil, as it can dissolve in the surface run-off water. The risk of such losses is enhanced in old grasslands, as these often have a turf layer on top with little phosphate binding capacity.

Plant residues left on the soil surface decompose and disintegrate to matter enriched in phosphorus. The process is accelerated by freezing and thawing and can lead to phosphate losses into surface water. Similarly, manure applied on the soil surface is susceptible to loss when heavy rains occur, especially when the ground is frozen. Some countries are regulating the spreading of manure to avoid such losses.

Phosphate can also be lost from poor storage and handling of manure.

The enviromental issues with losses of phosphate are discussed in later sections:

- eutrophication of waters, Section 4.8
- the sustainability of present use of mineral reserves of phospate rock, Section 4.12.

References: Aslyng (1988), Johnston (1989), Tisdale et al. (1985), Wild (1988).

4.5 The remaining nutrients: potassium, sulphur, magnesium, calcium and the micronutrients

A balanced supply of all mineral nutrients is required for proper crop development and yield. The health and performance of farm animals also depend on the correct amount and balance of mineral nutrients in feed crops. Both needs require attention.

Potassium *

Potassium is a common element. The earth's crust contains about 2.3 per cent potassium. It is one of the three major crop nutrients, with an essential role in physiological processes such as *water uptake*, *osmotic regulation*, photosynthesis and enzyme action. An adequate potassium supply is necessary for ensuring crop resistance to lodging, disease and drought.

The main soil sources of potassium are clay minerals. Much of soil potassium is present as part of insoluble mineral particles and inaccessible to plants. Only the slow process of weathering can liberate such potassium. Fertilization is required in order to ensure that crops get a sufficient supply of soluble potassium and that the soil reserves of accessible potassium are not depleted. Usual application rates are between 40 and 170 kg potassium per ha. Potassium binds to the surface of clay particles; this reduces leaching. But organic and light sandy soils can have little potassium binding capacity. Heavy application of potassium (notably with manures) on such soils can result in leaching losses. There are regions (e.g., in the Netherlands) where ground waters now have potassium concentrations above the maximum specified in the EC water directive. This states that drinking water should not contain more than 12 mg potassium/l, and preferably less than 10 mg.

However, the reason for this specification is that potassium can be an indicator of the presence of sewage which normally contains 20-50 mg potassium/l. Potassium in water at such concentrations has no detrimental health effect in itself; indeed it is an essential element for human nutrition.

* *Potassium is expressed as the element K. For conversion to K_2O, multiply by 1.2.*

For comparison, the potassium content of human milk is about 500 mg/l, and that of cow milk about 1400 mg/l. Normal human dietary intakes are:

▪ babies (2 months old)	about	800 mg potassium/d
▪ infants (1 year old)	"	1600 mg potassium/d
▪ adults	some	2000 - 6000 mg potassium/d

Source: (Danish National Agency of Environmental Protection (1984)).

WHO has not recommended a limit for potassium in drinking water. Potassium in water is not a limiting nutrient in eutrophication.

Potassium is weakly radioactive and is the main source of radioactive radiation from fertilizers. This topic is discussed in Section 4.6.

Sulphur

Sulphur is also an essential element for plant growth. Industrial emissions have up to now furnished sulphur maintenance dressings in many countries, through acidifying rain and aerial depositions. Reduced emissions may be one of the reasons why there is an increasing need in many parts of Europe for application of sulphur to maintain optimal conditions for plant growth. In some areas (e.g., Germany) concern is expressed about rising sulphate concentrations in ground waters believed to be caused by chemical reactions (e.g., denitrification) in deeper soil layers.

Calcium and magnesium

Calcium and magnesium are among the most common elements occurring in the earth's crust, but soils differ greatly both in content and the availability to plants of these nutrients. Mineral soils deficient in calcium are rare. Liming of the soil to correct acidity also contributes to maintenance of the soil's calcium levels. Maintaining the magnesium content of soil through application of minerals or fertilizers containing magnesium is a common practice. Grass forage can be deficient in magnesium if too much ammonium or potassium and too little magnesium is applied; such imbalances can cause the condition "grass tetany" in cattle or sheep.

Micronutrients

Micronutrients (chlorine, iron, manganese, zinc, copper, boron, molybdenum, also sodium, cobalt, vanadium and silicon) are usually present in soils in sufficient amounts, but deficiencies can occur. Excessive amounts are detrimental. Deficiencies, e.g., of boron and copper, can be corrected through application of special types of fertilizers with these elements added. Other micronutrients are often applied in crop sprays.

The mineral composition of plants depends on a variety of factors:

- parent material of the soil
- soil pH: increasing pH above about 6 - 7 decreases a plant's uptake of most micronutrients
- soil content of clay and organic matter
- moisture and aeration in the soil
- microbial activity with formation of organic compounds that combine with micronutrients and ease their uptake into roots
- relative abundance of ions in soil water
- plant species and variety which vary considerably in the uptake of trace elements.

Some crops have special requirements: sugar beet needs extra chloride, which would be detrimental to the quality of potatoes and strawberries by causing excessive water uptake.

As with all nutrients, the removal of the trace elements from the soil increases with increasing crop size and cropping can deplete soil reserves of trace elements to levels where replenishment becomes necessary.

Animals and man need some micronutrients not required by plants. For example, some regions (e.g., in Finland) have soil so poor in selenium that amelioration through fertilizers with added selenium is used to ensure a healthy diet. But there are also regions (e.g., in the USA) where the natural level of soil selenium is so high that herbage becomes poisonous to grazing animals.

It is suspected that some animal health problems, e.g., abortions, may originate in imbalances in essential micronutrients. The utilization of macro and micronutrients in man and animals depends on both their actual and relative content in the feed. As in plants, interactions occur. The availability of some minerals (phosphorus, zinc, etc.) may to some extent depend on the content of calcium in the diet.

A balanced supply of minerals is therefore essential and in practical animal feeding this is secured by the manufacture of standardized mineral mixtures. To meet the animals' requirements, mineral mixtures are often included in compound feeds.

Reference: Mertz (1986).

Claims are also made that plants require organic growth factors from soil. Such claims have not been substantiated and are at variance with much practical experience from vegetable production by water culture where soil is not present.

General references to this section: Marschner (1986), Mengel and Kirkby (1987), Tisdale et al. (1985), Wild (1988).

4.6 Other elements

Many elements do not appear to be necessary for plant, animal or human nutrition but nevertheless occur in the raw materials for fertilizer production. These elements are therefore found in small amounts in fertilizers. The greatest amounts are found in rock phosphate and so in fertilizers containing phosphorus. Potassium chloride and sulphate minerals have much lower contents of these elements and virtually none are found in many straight nitrogen fertilizers.

Table 4.6.1 lists the heavy metals found in phosphate rocks and compares typical contents with those found in soil. The contents in phosphate rock vary widely with the origin and type of rock.

Table 4.6.1 HEAVY METALS IN PHOSPHATE ROCK AND SOIL

Element	Average* mg/kg phosphorus	Range mg/kg rock	Average mg/kg Rock	Average mg/kg Soil**	Input to soil, 100 years fertilizer use mg/kg soil***
Arsenic	45	1 - 300	7	6	0.04
Cadmium	170	0.01 - 120	25	0.35	0.14
Chromium	1000	0.3 - 460	150	70	0.83
Cobalt	13	0.5 - 6	2	8	0.01
Copper	200	6 - 80	30	30	0.17
Lead	40	3 - 40	6	35	0.03
Manganese	200	6 - 300	30	1000	0.17
Mercury	0.2	0.01 - 0.10	0.03	0.06	0.0017
Molybdenum	33	1 - 10	5	1.2	0.03
Nickel	230	1 - 85	35	50	0.19
Zinc	660	3 - 800	100	90	0.55

* 15% P in rock
** Median
*** Topsoil (20 cm soil), soil volume weight 1.2 kg/l, application 20 kg P/ha • year for 100 years.

Source: Bowen (1979), Norsk Hydro.

In most fertilizer processes, some 60 to 95 per cent of the trace elements in phosphate rock becomes part of the product; the amount depends on the element and process.

Table 4.6.1 also gives the input of heavy metals to a topsoil with a century's use of phosphate fertilizer. It is apparent that this input is small compared with what is naturally present in an average soil, with the notable exception of cadmium.

Cadmium

The issues concerning cadmium in fertilizers are:

- soil accumulation of cadmium and the possibility of long term effects on crops and their quality
- plant uptake of cadmium and thus the cadmium content in the human diet
- damage to the soil's microbial processes.

Cadmium is highly toxic to man but less so to plants. Rock phosphate and thus phosphate fertilizers contain traces of cadmium. There is concern that the use of phosphate fertilizers will slowly increase the cadmium content of arable land, and that this might eventually give a cadmium level in agricultural produce above acceptable levels. The issue originates from a tragedy in Japan in the 1960s where poisoning (itai-itai disease) was associated with use of rice grown on fields irrigated with water containing cadmium from a mine. Human cadmium intake is mostly via food and tobacco; only a minor amount comes from drinking water and air pollution. Usually less than 5 per cent of the intake is absorbed, but this accumulates in the kidneys, liver and eventually in the bones. Smokers (20 cigarettes a day) will absorb about twice as much cadmium as non-smokers.

Average cadmium intake in the European diet is some 20 micrograms/day, while WHO recommends 70 micrograms/day as the upper limit. As population exposure is uneven, some people may be at risk even with the current average intake.

The mean cadmium content in the earth's crust is about 0.2 mg/kg. The median content of dry soil is about 0.35 mg/kg (range 0.01 - 2), of marine sediments 0.12 - 0.98 mg/kg and of seawater 0.02 - 0.25 mg/m^3.

Common cadmium compounds are rather volatile with boiling points in the range of 900-1000°C. Thus phosphate rocks of volcanic origin have a low cadmium content, and cadmium is emitted to air from volcanoes and from industrial heating of ores or other materials, e.g., steelmaking and garbage incineration. It is also present in sewage sludge (Hansen and Tjell (1981)). The use of sludge on arable land is therefore contentious and limits to cadmium levels in sludge that can be used for application are set in the EC (EC (1986)) and elsewhere.

Fertilizers can contain cadmium when sedimentary rock phosphate is used as raw material. Table 4.6.2 shows typical values for cadmium content in some important rock phosphates.

Table 4.6.2 PHOSPHORUS AND CADMIUM IN PHOSPHATE ROCKS

Type of rock	Phosphorus per cent	Cadmium mg Cd/kg rock	Cadmium mg Cd/kg P
Volcanic origin:			
Kola, USSR	17.2	0.15	0.9
Palfos, South Africa	17.2	0.15	0.9
Sedimentary origin:			
Bou Craa, Morocco	15.9	35	220
Togo	15.7	55	350
Youssofia, Morocco	14.6	40	274
Jordan	14.6	5	34
Texas Gulf, USA	14.4	40	278
Florida, USA	14.4	8	56
Negev, Israel	14.2	20	140
Khouribga, Morocco	14.2	16	113
Khneifiss, Syria	13.9	6	43
Gafsa, Tunisia	13.2	50	380

Source: Norsk Hydro.

The amount of cadmium and other trace metals can vary considerably within a deposit.

The introduction of cadmium to the environment is increasingly controlled by regulations. An EC directive (EC (1986)) sets an upper limit of 1 - 3 mg Cd/kg for arable soils. Regulations are also being proposed or enacted for maximum levels of cadmium in fertilizers.

It will take some hundreds of years with current fertilizer use before arable soils approach the proposed limits for cadmium content. However, slowly increasing the soil's cadmium content is undesirable. At present, the only way of making fertilizers with a low cadmium content is to use rocks with a low content, which would exclude important phosphate resources and make raw material supply very difficult. The fertilizer industry is developing processes for removing cadmium, so that fertilizers with low levels of cadmium should be available in the future from all rock resources. There are considerable technical difficulties but there is confidence that these will be overcome.

Cadmium input to cultivated land is from:

- acid rain and aerial depositions
- fertilizers.

Generally both inputs are equally important. Cadmium is lost from land or made inaccessible by:

- leaching
- removal with grazed or harvested crops
- bonding to and in soil minerals, especially on calcareous soils.

The accumulation and availability of cadmium in the soil depend on the balance of these processes, which varies with local conditions.

A crop's uptake of cadmium from the soil depends critically on soil factors.

Uptake generally increases with:

- increasing the soil's acidity
- " " " " cadmium content
- " " " " temperature
- decreasing " " binding capacity for cadmium
- " " " " humus content.

However, different soil treatments and nutrient regimes do not seem to affect a crop's uptake of cadmium markedly.

Fertilized, manured and untreated topsoils at Rothamsted, England, have shown increases in cadmium content of 27 - 55 per cent since 1850 from atmospheric deposition. In spite of this, there is little evidence of a long term increase in cadmium concentration in crops. The exception is a possible slight increase in wheat, a trend not seen in barley (Jones and Johnston (1989)). Similar results are reported from Sweden (Gunnarson (1983)) and Norway (Bærug et al. (1987, 1989)). The latter results illustrate these general findings (Table 4.6.3).

Bærug et al. have compared crops and soils on old fertilized arable land with those from newly broken land, formerly not fertilized. The cadmium content of the old arable land is some what higher than in newly broken soils, but with no corresponding increase in the cadmium content of the crops.

Table 4.6.3 CADMIUM IN SOILS AND CROPS, NORWAY

Comparisons between old and new cropland
(Include oats, grass and potatoes)

Area	Cd in soil, mg/kg		Cd in plants, mg/kg	
	Old soil	New soil	Old soil	New soil
East Norway	0.057	0.044	0.028	0.038
Mid Norway				
All samples	0.150	0.060	0.008	0.007
Organic soil	0.200	0.090		
Mineral soil	0.040	0.013		
South West Norway	0.098	0.091		
1st harvest			0.064	0.096
2nd harvest			0.045	0.072

Source: Bærug et al. (1987, 1989).

Plant uptake of cadmium and its accumulation in the parts used for food also depend on the plants themselves. There are major differences by species. Generally, cadmium content increases in the order:

cereal grains < root crops < leafy vegetables

Some studies indicate that plants can take up a considerable proportion of cadmium from aerial deposition on the leaves: as much as 40 per cent has been measured. This may be part of the reason for the large variation in cadmium uptake from place to place and from year to year.

The content of cadmium in crops therefore depends on many factors, and the cadmium content of fertilizers is not the dominant one.

The possibility that cadmium together with zinc, copper and nickel in combination can have a detrimental influence on the soil's microbial processes is a current field for investigation. Further work is needed to define critical input levels and their practical consequences.

Cadmium is thus definitely an undesirable element in fertilizers, though current fertilizer usage presents no immediate danger. The development of processes for cadmium removal should be given priority.

References: Anon (1989a), Krajnc et al. (1987), Ros and Slooff (1988).

Arsenic

Arsenic is not a rare element, its mean content in the earth's crust is 1.8 mg arsenic/kg. A few rock phosphates that are very low in cadmium contain 200 - 300 mg arsenic/kg. This has given rise to questions whether such phosphates should be used for fertilizer production.

Plant uptake of arsenic is known to be very low. Human diets should contain trace amounts of arsenic. Seafood is the main source.

Arsenic is bound to soil particles much as phosphate is. The arsenic content of dry virgin soil is generally less than 10 mg/kg. The use of arsenical pesticides has in some places caused topsoil concentrations of arsenic up to 600 mg/kg. Too much arsenic in the soil can be toxic to plants and reduce yields. Applications of about 300 - 30 000 g arsenic/ha can diminish plant growth. A single application of phosphate fertilizer made exclusively from phosphate rock rich in arsenic can add up to 60 g arsenic/ha. However, the presence of excess phosphate (as in a fertilizer granule) reduces the toxicity of arsenic to plants. Rocks with an arsenic content of less than about 15 mg arsenic/kg give little cause for concern. Such rocks dominate the world supply.

Fluorine

The element fluorine is highly reactive and occurs naturally only as fluoride. Excessive amounts of fluoride are toxic to most organisms, although fluoride is beneficial in small amounts. Human deficiencies are rare in developed countries. Where the water is fluoridated (1 mg F/l) the total diet may contain a mean of 2.7 mg F/day. Fluoride is essential for healthy bone and teeth formation, but too much causes mottled teeth and excessive amounts lead to severe bone diseases.

Fluorine is the thirteenth most abundant element in the earth's crust which has a mean content of 0.0625 per cent. Phosphate rock contains 3 - 4 per cent fluoride. Between 50 and 80 per cent of the fluoride content in phosphate rock remains in the fertilizers.

Fluoride in soils is mostly present as insoluble soil minerals. It is removed from soil solution by calcium and aluminium, e.g., through formation of the mineral calcium fluoride. Clay particles bind fluoride. Increasing soil acidity can increase the fluoride concentration in soil water but fluoride is not readily taken up by plant roots, as shown by crops growing on soils with a high fluoride content. Thus it is unlikely that fluoride application associated with phosphorus fertilization will affect the crops. Fluoride leaching from most soils is insignificant.

A very different subject is the toxic action of hydrofluoric acid on plant leaves; plants are generally highly susceptible. Industrial air emissions of fluoride, e.g., from aluminium smelters, are thus severely controlled.

Fertilizer influence on plant uptake of aluminium

Claims that fertilizers and especially nitrate increase the solubility and plant uptake of aluminium have been refuted by field measurements (Åkerstrand et al. (1988)). Most plant roots exclude aluminium efficiently, and the control of soil pH through liming limits root exposure to dissolved aluminium.

Radioactive elements

The human body is exposed to radiation from several sources:

- about 90 per cent of the dose comes from external sources: cosmic radiation and radioactive elements in soil, buildings, etc.

- some 10 per cent is due to internal sources: radioactive elements in the body itself.

Potassium is weakly radioactive. The isotope K-40 comprises 0.0118 per cent of the element. K-40 emits ß-particles and has a half-life of 1.8×10^9 years. Potassium is an essential element for all living creatures, the adult human body contains 2 g potassium/kg. The body content of K-40 is therefore equivalent to 60 Becquerels (Bq)/kg, providing some 9 per cent of the radiation from internal sources.

The amount of K-40 in the soil varies between 200 and 2000 Bq per kg topsoil and on average about 20 per cent of human exposure to radiation from external sources is due to K-40 in the soil.

Fertilization with potassium replenishes soil losses of this element and does not change the amount of environmental radiation from K-40.

Rock phosphate contains some uranium, thorium, radium and their daughter products, all radioactive elements. The content of these is about 5 - 50 times higher in phosphate fertilizers than in normal topsoil. The fertilizer content varies greatly depending on rock source, ranging from 1700 to 9000 Bq/kg fertilizer. Most of this radiation comes from uranium and radium.

The elements uranium and radium and their radiation are naturally omnipresent. They contribute about 80 per cent of human exposure to radiation from internal sources and some 30 per cent of that from external sources. Normal use of phosphate fertilizers can, over time, increase the soil content of these elements and consequently human exposure, but as a whole this increase will be very small compared to the great variation in soil content of the radioactive elements uranium, thorium and radium.

Reference: Ormberg (1987, 1988).

Other trace elements

Phosphate rock contains many other of the elements in the periodic system, but mostly in such small quantities that they are insignificant as sources of essential elements for crops and animals, or as soil contaminants.

General references: Adriano (1986), Mengel and Kirkby (1987), Wild (1988).

4.7 Erosion and fertilizer use

Soil erosion is the removal of soil from the land through the action of wind or water. It is a natural process that occurs even without human intervention. However, most forms of agriculture increase the erosion potential, especially practices that leave the surface of erodible land unprotected. Excessive erosion is a matter for serious concern. In a sustainable agriculture, soil erosion should not exceed the slow process of soil formation, and the prevention of soil erosion is a key issue in increasing the sustainability of agriculture.

Excessive erosion occurs with large variations in extent and causes between and within regions. It is difficult to measure and evaluate the gravity of the problem, but erosion is of special concern in areas such as the humid tropics, along the deserts and in parts of North America. In Europe, erosion is most serious in the Mediterranean regions.

Heavy erosion can remove fertile soil. Permanent plant cover is necessary in such land depressions to protect against erosion. Photo: B. Rognerud, Center for Soil and Environmental Research, Norway.

Soil erosion by water generally begins where raindrops strike bare soil. Soil aggregates are broken up, the surface compacted, and water infiltration into the soil obstructed. Water with suspended fine soil particles runs off as surface water, giving sheet erosion, where a thin layer of surface soil is removed. The water flowing over the soil surface can form networks of eroding channels that cut into the topsoil. In the worst cases deep gullies are formed. Suspended particles increase the water density and channelling increases the velocity of water flow. Consequently, erosion starts gently and then rapidly accelerates.

The removal of forests has reduced water infiltration into soil in catchment areas and increased flood frequency and destructiveness. Floods enhance erosion.

Eroded material eventually settles out, filling up water reservoirs and estuaries. The silt deposit can improve the fertility of the receiving area, but in general soil erosion degrades agricultural land.

Wind erosion occurs when bare soil is exposed to drought and wind, e.g., the dust bowl in the USA in the thirties and more recently in the USSR.

It follows from the mechanism of erosion that:

- sloping land is at greater risk than flat land, sloping land left fallow during the winter is at special risk
- erosion risks vary with soil type and structure
- vegetation reduces erosion, as leaves intercept raindrops and roots prevent channelling.

Overgrazing has damaged fragile grasslands and caused serious erosion, e.g., in Africa. Ploughed land is at greater risk to erosion than grasslands. Specialized arable cropping generally suffers more erosion than mixed farming because with mixed farming part of the land is under grass and more organic matter is available for return to the arable part of the land. This gives some protection against erosion.

The extent of erosion is greatly influenced by soil management. As an example, Table 4.7.1 gives erosion losses from two Norwegian plots with identical slope but different soil type under varying operations:

Table 4.7.1 EROSION LOSSES UNDER DIFFERENT MANAGEMENT

Soil loss in different agricultural systems when the risk of erosion is high. Slope 1:8. Mean values for 1982-86.

	Soil loss, tonnes/ha	
System	Ås, loam	Øsaker, clay
Fallow	35.9	3.2
Spring barley autumn ploughed, spring harrowed	9.3	1.0
Spring barley, not ploughed, spring harrowed	2.7	0.6
Meadow	0.2	0.02

Source: Skøien (1988).

Techniques are available for reducing soil erosion, e.g.,

- water interception with soil banks, strips of grass or forests
- contour ploughing
- use of winter or catch crops, intercropping
- mulching
- no-till practices
- ditching
- terracing, forming horizontal patches of land on steep hills, a characteristic man-made landscape feature both in South-East Asia and elsewhere.

Proper fertilizer use can help minimize erosion by ensuring an ample supply of roots and plant residues. Where erosion has removed topsoil, liming and fertilization help the reestablishment of a good plant cover. Practices that increase erosion such as continuous arable cropping on land prone to erosion are made economical and possible by fertilizer use. But fertilizer use as such does not increase erosion.

References: CAST (1982), Schwertmann et al. (1989).

Erosion protection in Malaysia. Terraced fields with a variety of crops.
Photo: T. Bølstad, Samfoto, Oslo.

4.8 Eutrophication of fresh and marine waters

The available amount of nutrients is a key factor for species composition and biological production in water. An increasing abundance of nutrients in water leads to increasing production, a process called eutrophication. Moderate increases may have beneficial effects, e.g., increased fish stocks. However, if the primary production becomes too high or toxic algae form blooms, the result will be regarded as detrimental.

Phosphorus, nitrogen, silicon and a number of other elements such as cobalt, molybdenum and iron are essential as growth factors for algae and can be growth regulating factors. Some types of algae utilize organic matter, e.g., from sewage. Some accumulate nutrients for later use. Phosphorus and/or nitrogen are the main factors determining growth. Hence eutrophication of water is usually due to excess input of these two nutrients.

The primary effect of eutrophication is an increase in algal biomass and the relative abundance of species can also be altered. The species composition and abundance also depend on the stocks of predators and grazing efficiency. With eutrophication changes may occur also at higher levels in the food web (e.g., fish). With increased frequency of algal blooms, the probability increases that a toxic bloom will occur.

The increase in algal biomass due to eutrophication requires a correspondingly increased supply of oxygen for decomposition of the organic material. When the algae and their remnants sink to the bottom, reduced oxygen content and eventually anaerobic conditions may result. This is one of the symptoms of severe eutrophication.

In lakes and deep fjords oxygen deficient conditions at the bottom may be a natural phenomenon due to poor water exchange, and are not necessarily a result of pollution. A more detailed discussion of aquatic ecosystems and of eutrophication in fresh and marine waters is found in Barnes and Mann (1980), R.B. Clark (1989) and in Olsen and Jensen (1989).

Fresh water

Eutrophication in lakes has been thoroughly studied. Useful empirical models predicting the effect of nutrient inputs have been established as tools in water quality management and the evaluation of measures for reversing eutrophication.

In most cases, phosphorus is the nutrient which limits the growth of algae. In natural drainage water from land areas unaffected by human activity, phosphorus is the least available element in relation to the needs of the algae. The normal nitrogen/phosphorus ratio (based on weight) for optimal algal growth is about 7 (species dependent range 4 - 10). In natural waters this ratio is around 20/1 or more. Even in waters receiving large sewage inputs with a nitrogen/phosphorus ratio of 4 - 5, phosphorus can be limiting in the sense that a reduction in phosphorus input has visible effects on the eutrophication level.

When the nitrogen/phosphorus ratio is less than 4 - 5, nitrogen can be the growth limiting nutrient. As it is easier to control phosphate than nitrogen emissions, the most cost effective way to control eutrophication is phosphate removal from waste waters by proven technology.

A low nitrogen/phosphorus ratio in lakes can sometimes give nitrogen fixing blue-green algae (Cyanobacteria) a competitive advantage over other species. In such cases, nitrogen from biological nitrogen fixation becomes available so that the algae can utilize the growth potential that the phosphorus represents.

A nitrogen/phosphorus ratio which favours blue-green algae is considered unfavorable because:

- blue-green algae may be toxin-producers and give unpleasant odours
- blue-green algae are not well utilized as food for zoo-plankton. The ungrazed mass increases the risk of oxygen depletion when it decomposes
- blue-green algae create more aesthetic problems than other types of algae.

Marine environment: The Baltic and the North Sea

Serious pollution and eutrophication now occur in the Baltic, the Kattegat, the South-Eastern parts of the North Sea and in many of the adjoining coastal and fjord areas. The present situation is described by Aniansson (1989), Hognestad (1987) and the UK Department of the Environment (1987). Figure 4.8.1 indicates the main affected areas.

Figure 4.8.1 REGIONS IN THE BALTIC - KATTEGAT - NORTH SEA REPORTED TO BE AFFECTED BY EUTROPHICATION

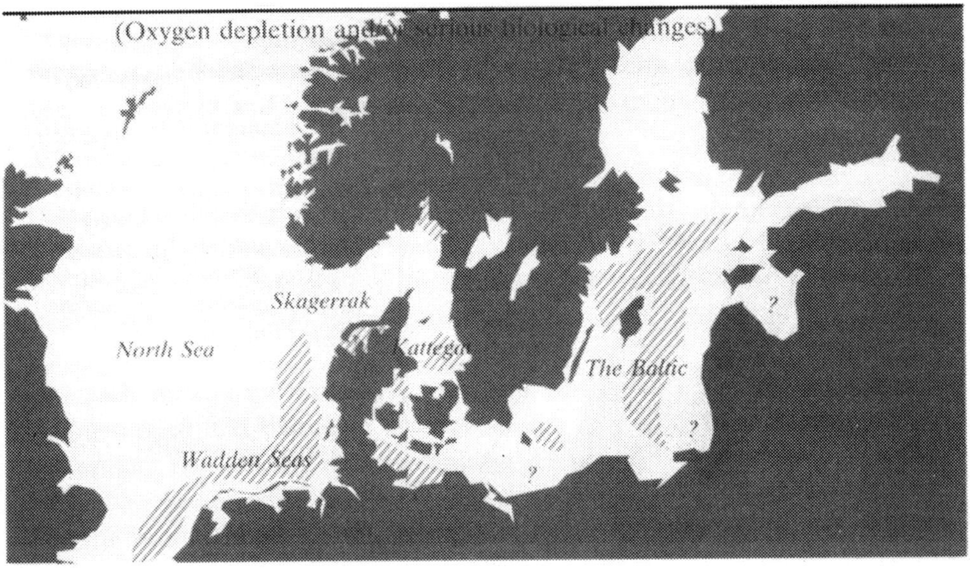

Most of the North Sea is oxygen saturated and not greatly affected by eutrophication, though there have been increases in the amounts and changes in the species composition of plankton algal stocks in recent years. But in the Dutch, German and Danish coastal regions, signs of eutrophication are clearly present. Here the currents transport river water with nutrients along the coast. The British coast is little affected in spite of large nutrient inputs from rivers flowing into the North Sea because the tides promote rapid water exchange.

The North Sea is not a unique case. There is also serious eutrophication in other marine areas, e.g., in some Mediterranean and United States coastal regions.

Increased fish catches during the past decades may have been due to increased nutrient input to the sea. But the optimal production level seems to have been passed, as indicated by the extended areas with oxygen deficiency at the bottom in the Baltic, Kattegat and the Wadden Seas.

The nutrient input to the Baltic has increased above a natural level by a factor of about 4 for nitrogen and 8 for phosphorus. For Kattegat-Skagerrak the corresponding factors are 6 and 10 - 20.

It has been estimated that the combined input from the Scandinavian countries and atmospheric deposition to the Skagerrak has a nitrogen-phosphorus ratio of 27 and for the Kattegat 20 (Hognestad (1987)), for the

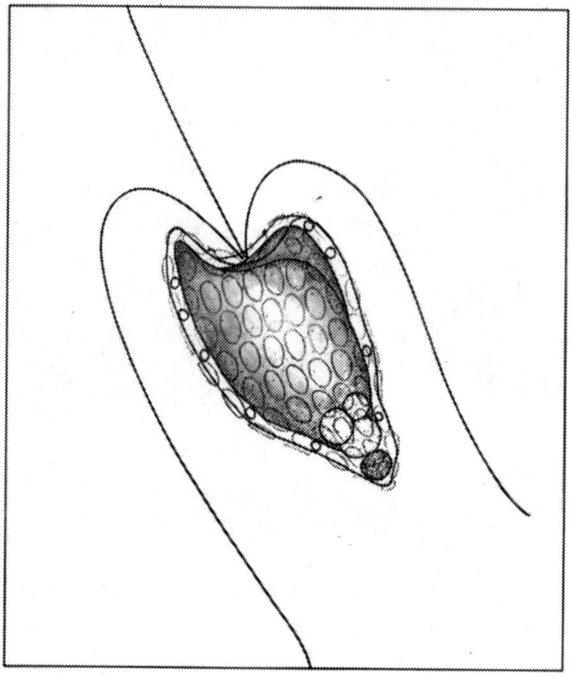

The alga Chrysochromolina polylepis.

The bloom in 1988 of a toxic strain of this alga killed fish and other marine animals along sections of Danish, Norwegian and Swedish coasts. This event focused public attention on marine pollution.

Drawing: J. Throndsen, Institute of Biology, University of Oslo.

Baltic about 16 and the North Sea south of 56°N about 10 (Gerlach (1987)), while the normal seawater ratio is about 7. The dissolved nitrogen is mainly present as nitrate because conversion of ammonia to nitrate is a fairly rapid process in water.

In the North Sea as a whole, 83 per cent of the nitrogen and 87 per cent of the phosphorus come naturally with the flow of Atlantic water. Nutrients from human activities come mostly with river water and land drainage and affect mainly coastal regions. This input has increased by a factor of 4 for nitrogen and 7 for phosphorus above the natural level. The relative contribution from agricultural, industrial and urban sources is uncertain, but Lidgate (1987) estimated that of the nutrients coming from human activities, about 60 per cent of the nitrogen and 25 per cent of the phosphorus might originate from agriculture. The rest comes mainly with sewage. In addition to the water-born nutrients, both the Baltic and the North Sea get a substantial input from atmospheric deposition. This accounts for about 20 - 30 per cent of the land-based nitrogen input to these waters. Some of this also originates from agriculture (Section 4.3).

The marine environment is not uniform. The Baltic-North Sea regions consist of fjords and sounds, brackish water, coastal areas, shallow and deep areas, and the water masses are stratified in distinct layers. It is thus an interconnected but very complicated system, with differing and changing rates of nutrient and oxygen supply, two of the many factors that determine biological growth.

The complexity of the system makes it difficult to interpret field observations and determine what factor is growth limiting in the various areas and water bodies; it may not be the same factor in all places, e.g., in coastal areas and the open sea, and it may change with the seasons.

Research on marine eutrophication has not established clear connections between load and response similar to that in fresh water. At present, two contending theories are debated among scientists.

The traditional view is that nitrogen is the main limiting nutrient for algal growth in the sea because laboratory experiments indicate that the addition of nitrogen to sea waters increases growth (Howarth (1988), Larsson (1988)). This is especially so for the coastal waters where indications of nitrogen limitation are often found.

The other view accepts these observations, but points out that such data often refer to special local conditions and that this does not validate the reverse conclusion: that reduction in nitrogen input to the sea will be an efficient general measure for eutrophication control (Söderström (1988), Sakshaug (1988)).

One reason for this is that the increased supply of decomposing organic matter in eutrophied waters favours denitrification. This results in a loss of nitrogen while there is no similar rapid removal process for phosphates. Hence denitrification should make the nutrient balance insensitive to a reduction in nitrogen input. The deficiency of nitrogen observed in experiments and field observations can be a result of this natural process. Denitrification may reduce the nitrogen availability to about the level optimal for algal growth at the phosphate level present. Thus the primary production in eutrophic areas could be determined mainly by phosphorus availability.

In this unresolved situation, the authorities (e.g., in Denmark) demand reduction both in nitrogen and phosphorus inputs.

Denitrification in the marine environment is incompletely understood at present and details about the extent and speed of the process in the

Kattegat/North Sea coastal area are largely lacking. Even less is known about the possible contribution from nitrogen fixation taking place in the North Sea. This process is estimated to be of little significance in sea water, but in the brackish waters of the Baltic it may provide about 15 per cent of the total input of combined nitrogen (Andersen (1986)).

Phosphorus does not undergo any parallel conversion cycles. Phosphorus is eliminated through sedimentation of particles and incorporation in the sediments.

4.9 Fertilizers and weeds, plant diseases and insect problems

In this section, the word pest will be used for weeds, plant diseases and insects detrimental to crops. The control of these has always been an important part of agriculture, as discussed in Section 2.3. The supply of plant nutrient is one of the factors influencing the incidence of pest problems in agriculture and the damage done to crops. Weather, crop variety, previous cropping, tillage practices and presence of infection sources are also of major importance.

Weeds compete with crop plants for nutrients, water and sunlight. Weed infestation must therefore be kept below tolerable limits. Ample nutrient supplies benefit both the weed and the crop. Fertilization increases early growth rate and canopy closing and thus gives the crop a competitive advantage over some weeds, but also furthers the growth of tall fast-growing weeds, especially perennials. Some recent changes in agricultural practice (increased use of winter cereals, elimination of straw and stubble burning) have increased weed problems. The main control measures are chemical weed-killers (herbicides) in addition to traditional mechanical methods such as harrowing and ploughing that disrupt and bury the weeds. When tillage is used sparsely (e.g., to reduce soil erosion), the need for herbicides is increased.

After weeds, diseases are the most serious and difficult pest problem in agriculture. The causes are fungi, viruses and bacteria, with fungal diseases being the most important.

Fungicides are used both on seeds and crops. However, fungi can develop resistance, and protective agents are not available for some important diseases, e.g., take-all in wheat. The registration and use of pesticides are controlled by regulations (Council of Europe, (1984)).

Intensive agriculture has increased crop susceptibility to disease, for a number of reasons.

Fertilization favours the growth of the crop. This can in turn favour the growth and reproduction of parasites, and increase the need for crop protection, notably by fungicides. Lush growth leading to an early closing of the crop canopy makes for humid conditions within the crop, which benefits the fungi.

Attacks of some root diseases are enhanced, others are impaired, by increasing soil acidity and by the ratio of ammonium to nitrate in the soil. Fertilizers can thus affect such diseases through their (mostly) acidifying action.

Plants make protective chemicals and barriers which restrict the progress of an infection, and compensate for the damage through new growth. The ability to mount such metabolic defences depends, among other factors, on a sufficient nutrient supply. Adequate potassium supply is of notable importance. Lack of nutrients as well as an unbalanced supply of nutrients such as an excessive application of nitrogen can distort plant development and predispose to disease.

The connection between fertilization and plant diseases is thus complex and varies both with the plant and the pathogen, but in general excessive nitrogen levels increase a plant's susceptibility to disease, as do potassium deficiencies. The effect of phosphorus is variable. Micronutrient deficiency (e.g., boron) can predispose plants to disease. Properly timed and balanced fertilization is a supporting measure in crop protection, whereas excesses can be detrimental.

Other factors in intensive agriculture not directly related to fertilizer use enhance the susceptibility of crops to disease. Large continuous areas planted with the same crop increase the possibilities for the rapid spread of disease, and the modern high-yielding varieties are so closely related that there is some risk that new diseases may meet little resistance from genetic variation. Thus plant disease control is of paramount importance.

Some control of disease is obtained by breeding resistant varieties of plants and by using healthy seeds. Liming prevents clubroot on crucifers, but can increase take-all attacks on cereals. Plant diseases are mostly crop specific and crop rotation prevents the transfer of infection from one crop to the next. But as cereals are grown on more than half of the arable land in Europe, the possibility of crop rotation is limited in specialized arable cropping. Organisms which suppress pathogens can accumulate in permanent cereal fields, but this may take some years.

Insects damage plants both by feeding and by acting as disease carriers. Thus Dutch elm disease is spread by a bark beetle. Insect populations depend on complex interactions between plants, insects and predators, and the supply of plant nutrient is only one of the factors involved. Nitrogen is a limiting nutrient for insects as for plants, and fertilized fields carry increased biomass. Fertilization furthers dense crops with a humid microclimate which benefits some insects. Thus fertilization can promote insect attacks on crops, although there are conflicting reports and the connection is uncertain.

References: Agrios (1988), Engelhard (1989), Hauch (1984), Huber and Arny (1985).

4.10 Plant nutrients and produce quality

The concept of quality as related to agricultural produce is highly complex.

Quality refers principally to:

- visible or outer quality: shape, size, colour, absence of blemishes, etc.
- inner quality: criteria important for assessment of nutritional value, suitability for transport and storage, technical processability, taste and odour.

Examples of quality parameters are:

- for dietary value: protein, fat, starch, vitamins, etc.
- for technical processability: baking and brewing quality, sugar content in sugar beet, etc.
- for storage and transportation: water and sugar content, firmness.

As regards chemical quality, criteria exist for the main crops: cereals, sugar beet, oil crops, potatoes and herbage for animal feeds. The influence of nutrient supply on such quality parameters forms part of this section.

Claims that alternative agricultural crop production systems give produce with superior quality are discussed in Section 5.5.

Produce quality and fertilizers

Nutrients

The various nutrients (and lack of them) have a variety of influences on those biochemical processes that eventually determine the quality of produce. Nutrient supply and balance are therefore important factors in determining produce quality as well as yield.

The supply of nitrogen determines a plant's growth, vigour and yield, colour and protein content. As the supply of nitrogen increases, a crop's content of protein and some of the vitamins A and B tends to increase and that of vitamin C to decrease (Hög (1985a)). The sugar content decreases and skins and cell walls become thinner, notably when the supply of nitrogen is excessive in relation to other nutrients. The amount, form and timing of nitrogen application have thus a major influence on produce quality.

Sufficient phosphorus is required for good rooting and resistance to drought, for plant growth and development, and for ripening of seed and fruit. These are all factors influencing the final quality of the produce.

Sufficient potassium is important for vitamin and mineral content, for texture, firmness and resistance to transport damage.

Insufficient nitrogen in relation to phosphorus supply can cause premature ripening. This can influence quality, e.g., by increasing the content of dry matter. Excessive nitrogen induces delayed ripening and increased water content, especially if the weather is wet at harvest.

The mineral content and composition of plants is influenced by the nutrient balance, e.g., excessive potassium can reduce the uptake of calcium and magnesium.

Lack of trace minerals can result in both reduced yield and a variety of quality faults. Quality thus also depends on soil factors such as acidity. Grass grown on acid soils can be deficient in cobalt, an essential element for animals.

Optimal fertilization, with a correct balance between nutrients, is required to ensure both high yield and high quality. The nutrient supply giving the optimal quality for the desired use is not always identical with that giving the maximum yield, although it is not necessarily different.

Recommendations on the use of fertilizers take such matters into consideration. Incorrect use of fertilizer or manure can give quality problems, but such problems should not be exaggerated or taken as typical of the relationship between produce quality and fertilizer use.

Examples of crop qualities determined by the availability of mineral nutrient are:

Wheat:
The baking quality of wheat flour is dependent on the protein content. The amount and quality of the protein gluten is especially important. An EC criterion for premium wheat is that it should contain at least 14 per cent protein. This requires a sufficient supply of nitrogen often including a top-dressing of late nitrogen.

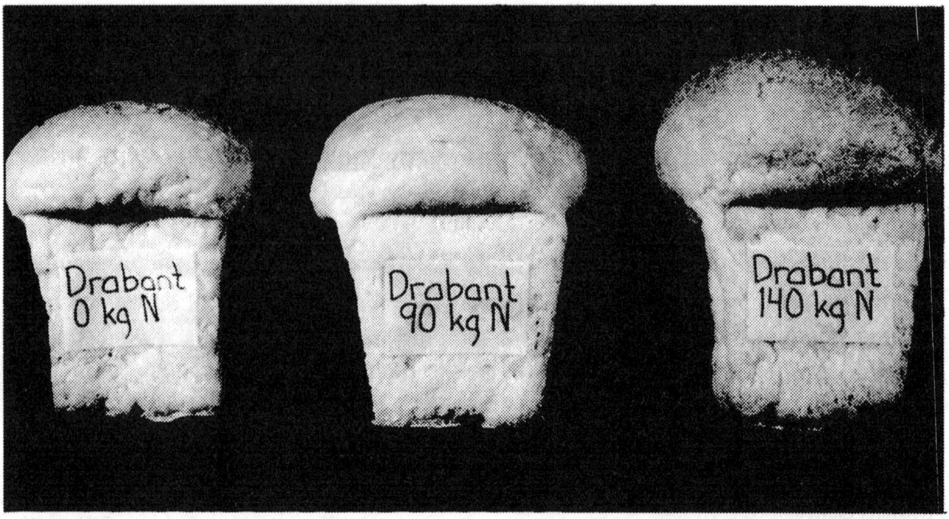

Three different nitrogen application rates on spring wheat in Sweden demonstrate the effect of nitrogen fertilization on the baking properties of wheat flour. Photo: Hydro Supra.

Barley:
For feed purposes barley should have sufficient nitrogen to ensure a high protein content while barley for malting should preferably have a low protein content. Thus the end use can influence the amount of fertilizer that should be used.

Total crop protein is a mixture of many proteins differing in composition, and some increase more than others with nitrogen supply. This can be seen from the content of the essential amino acid lysine. The grain content of lysine increases with increasing nitrogen supply, but its relative content in the total protein decreases.

Sugar beet, potatoes and oil crops:
The amount of sugar and impurities in sugar beet are strongly influenced by nitrogen supply. Excessive nitrogen reduces sugar content and can result in lower prices for the beet.

In oilseed rape, oil content is influenced by the supply of nitrogen and the content of glucosinolates (noxious natural organic sulphur compounds) can be affected by the supply of sulphur.

In potatoes, the content of starch and dry matter in the tubers depends markedly on nutrient supply. Increasing nitrogen application usually causes a drop in starch and dry matter content in the tubers. Padmos (1986), and Prummel (1981) found that the starch content was optimal with a potassium application of 75 kg K/ha. It then decreases with higher applications.

These numbers illustrate the effects that can be found. The magnitude of the effects and the application rates related to them varies with the soil type, climate and variety. For all these crops, fertilizer use can be adjusted to take account of quality as well as yield.

References: CEA, IFA, IPI (1983), Hauch (1988), Hög (1985a), Mengel (1979).

4.11 Energy use in agriculture

Farmwork and energy

Energy use on farms is necessary to make cropping possible. The work is often heavy:

- soil tillage
- sowing
- control of weeds and other pests
- fertilization
- drainage and irrigation
- harvesting
- animal care
- handling of manure
- transport.

Buildings, machinery, fertilizers, pesticides, irrigation and drainage equipment etc. all require energy for their manufacture and maintenance.

The use of energy has greatly enhanced productivity of farmwork.

Table 4.11.1 MANHOURS REQUIRED FOR TILLING 1 HA OF SOIL

	Hours
Manpower	400
Oxen (pair)	65
6 hp tractor	25
50 " "	4

Source: Pimentel and Pimentel (1979).

Fossil fuel and machinery have largely replaced human toil and animal power in the developed nations, with many additional benefits, e.g.,

- some 10 per cent of Europe's agricultural land was used to maintain draught animals, which is now used for other purposes, e.g., cattle
- the growing season has been extended as tilling, sowing and harvesting times have been reduced
- yields have increased due to fertilizers and pesticides
- crop drying ensures storage without spoilage.

But these benefits are all currently based on the use of non-renewable resources, with a few exceptions, such as hydro-electric power.

Use of non-renewable energy in agriculture

The use of energy from fossil fuel in agriculture (including the production and use of fertilizer) should be seen in the context of the use of non-renewable sources of energy in modern society as a whole.

Agriculture's principal business is food production. Other sectors (e.g., fibre) are minor in comparison.

Various estimates have been made of the energy used in the food sector. The conclusions are similar.

Table 4.11.2 ENERGY USE IN THE FOOD SECTOR

(Estimates of energy use in per cent of total national energy use)

Activity		Per cent
Agriculture		5
Food processing		5
Transport, storage, sales		1-2
Home/restaurant preparation		3-4
Food sector	about	15

The remaining 85 per cent of the energy use in society is industrial (excluding food and fertilizer production), for transport, public and residential (non-food) use and for military purposes.

The energy used in agriculture (5 per cent of total energy use) can be broken down among processes, Table 4.11.3.

Table 4.11.3 ENERGY USE IN AGRICULTURE

Process	Per cent, estimated
Fertilizer production	40-60
Machinery, fuel, etc.	40-60
Irrigation	2-10
Pesticide production	2

Fertilizer production currently accounts for about 2 - 3 per cent of total energy consumption. The major item is ammonia production. Energy consumption in modern fertilizer factories is given in Table 4.11.4.

Table 4.11.4 ENERGY USE IN FERTILIZER PRODUCTION IN MODERN FACTORIES

Fertilizer	Energy consumption		
Nitrogen fertilizers:			
Ammonia	35.3	MJ/kg	nitrogen
Urea	42.3	"	"
Ammonium nitrate	34.8	"	"
Phosphorus *	12-19 **	MJ/kg	phosphorus
Potassium Chloride *	5	MJ/kg	potassium

* Including energy used in mines.
** Depending on product and process.

Source: Norsk Hydro.

For old factories, the total energy use can be more than 50 per cent higher.

Fertilizer production has become more energy efficient through improved design. Ammonia factories built now (1990) use some 30 per cent less energy per tonne of nitrogen than those designed around 1970. Present design is thermally very efficient, and further energy saving beyond another 5 - 10 per cent is not feasible in ammonia production. Other potential nitrogen fixing methods (e.g., high temperature oxidation) are much less energy efficient.

The presently preferred energy source in fertilizer production is natural gas. Hydroelectric power can also be used, though this would require higher energy consumption and major rebuilding of the factories.

As energy input comes mainly with the applied nutrients and the fieldwork, it varies greatly with crops. Table 4.11.5 gives an estimate of the energy input for growing barley in Norway, and illustrates the relative importance of different items of energy input in cereal growing.

Table 4.11.5 ENERGY INPUT FOR GROWING BARLEY ON 1 HA IN NORWAY

(Net yield 3670 kg/ha)					Energy input	Energy input subtotals
Fertilizers and agrochemicals:					MJ/ha	MJ/ha
Nitrogen	100	kg	à	35.3 MJ/kg	3530	
Phosphorus	33	"	"	16.0 "	528	
Potassium	60	"	"	5.0 "	300	
Lime	70	"	"	1.0 "	700	
Agrochemicals	25	"	"	111.0 "	278	
Transport and packaging					976	6312
Field work:				fuel l/ha		
Ploughing				23.3	941	
Levelling				4.7	190	
Harrowing (3 times)				14.0	565	
Sowing and fertilization				5.2	210	
Rolling and spraying				5.2	210	
Harvesting				20.7	836	
Farm transport, maintenance				9.6	388	3340
Crop cleaning and drying (ambient air)						414
Investments:						
Machinery					3103	
Ditches	1250 m à 1.15 MJ				1438	
Buildings					577	5118
Energy input total:						15184

Based on Breirem et al. (1984) and data from Table 4.11.4.

Pimentel and Pimentel (1979) give similar accounts for a wide variety of agricultural systems. The energy input in much of agriculture in developing nations is only a fraction (somc 10 per cent) of that in the developed nations. Where unemployment is rife and labour cheap, increased energy input from extensive mechanization can be inappropriate. But increased energy input through use of fertilizers is necessary for increased food production in the developing nations. The development of local energy sources (gas fields, hydro-electric power) is therefore part of their struggle to safeguard and increase their agricultural productivity.

Energy efficiency in agriculture

The principal source of energy in crop production is the sun: the plants capture and use the sun's energy, and this energy is harvested in the form of a crop. Farmers invest an energy input (labour, fuel, fertilizer, etc.) in order to obtain the crop. The ratio between harvested energy and energy input depends on the agronomic system. Pimentel and Pimentel (1979) cite many such energy ratios from various crops (Table 4.11.6). Generally:

- the growing of cereal and root crops gives a net energy gain
- intensive horticulture (fruit and vegetables) produces less energy than the input
- intensive animal rearing consumes more energy than it delivers as eggs, meat and milk; the losses of nitrogen referred to in section 4.3 reflect this
- traditional agriculture in developing nations appears more energy-efficient than current agriculture in industrialized countries.

Table 4.11.6 NET ENERGY GAIN AND LOSS IN AGRICULTURE AND HORTICULTURE IN DEVELOPED NATIONS

Ratio: Harvested energy/energy input

Crops with more energy gained than input	Ratio	Crops with less energy gained than input	Ratio
Maize	2.34 - 2.93	Brussels sprouts	0.69
Wheat	2.41 - 3.51	Spinach	0.23
Rice	1.55 - 2.45	Tomatoes	0.60
Oats	2.4 - 3.11	Oranges	0.37
Sugar beet	3.62	Apples	0.53
Potatoes	1.57		

Data from Pimentel and Pimentel (1979).

However, important as such calculations are for an insight into energy flow in agriculture, there are also other considerations that seem more relevant. In agriculture, chemical energy is not only used; it is transformed. We eat bread, oranges, meat and cheese, not coal and natural gas. Animal production can be wasteful, but much land is only suitable for grass production. Energy waste should be avoided, but properly used, energy inputs into agriculture are among the most rational ways of using this resource.

References: Breirem et al. (1984), Helsel (1987), Jürgens-Gschwind and Albrod (1981), Mudahar and Hignett (1982), Pimentel and Pimentel (1979).

4.12 Resources

It is held against industrialized society and its current agriculture that the high productivity is due to the use of exhaustible resources, that it is thus not sustainable in the long term, and that it will face a resource crisis sometime in the future. This is true for many aspects of modern society. The use of limited non-renewable resources, e.g.,

- coal, oil and natural gas
- mineral deposits

is central to the topic of sustainability.

It is usual to differentiate between reserves and resources, Figure 4.12.1.

Figure 4.12.1. THE RELATIONSHIP BETWEEN THE CONCEPTS "MINERAL RESERVES" AND VARIOUS TYPES OF RESOURCES

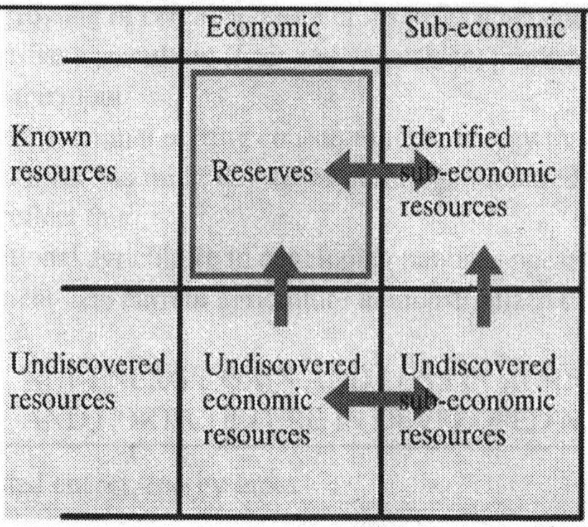

Reserves are well known and charted mineral deposits which can be worked with profit by present technology. Resources are minerals which may be made available in the future, through geological discoveries or technical or economic developments.

This section is mainly based on reserves as the most precise concept.

The time span that reserves will last given current consumption can give the impression that scarcity will appear at an earlier date than if resources are also taken into consideration. But at some unknown time in the future, the present abundance of fossil fuel and other mineral resources should eventually turn into shortage.

Availability of energy is a critical factor in the feasibility of utilizing dilute or remote resources. Their use will undoubtedly change under economic and political pressures in the future, just as they have done in the past.

Energy

Current global reserves of fossil fuel are given in Table 4.12.1, and present consumption is given in Table 4.12.2.

Table 4.12.1 GLOBAL ENERGY RESERVES

Reserves of	1000 million tonnes	Years before exhaustion at current consumption
Coal	1023	222
Oil	124	41
Natural gas	91*	58

* Calculated as oil equivalents.

Source: BP (1989).

Table 4.12.2 GLOBAL ENERGY CONSUMPTION 1988

Consumption of	Million tonnes of oil equivalent *
Oil	3000
Coal	2428
Natural gas	1631
Hydroelectric power	537
Nuclear	439

* Calculated as the amount of oil required to give the same amount of heat or produce the same amount of electricity.

Source: BP (1989).

Other resources (e.g., nuclear and geothermal energy; wind, wave and solar power) are being developed. Hydroelectric power at present represents about 6 per cent of global energy use. There is potential for expanding this about 6 times, mostly in the developing nations. Hydroelectric power is a renewable energy source without emissions, but there are other environmental issues involved, e.g., flooding above dams.

Efficiency and economy in energy use are a prerequisite for the increased sustainability of our society. As described in Section 4.11, agriculture should make its contribution to this end, but the major challenges are in other sectors of the economy.

Resources of raw materials for fertilizers

Fertilizer production and use depend on mineral deposits. Reserves and known resources of these deposits should in most cases last a few hundred years or more. It is expected that further resources will be discovered, extending the time horizon.

The agricultural use of micronutrients is only a fraction of the current consumption of these elements. Substitutes for their industrial use would greatly increase the sustainability of these resources because there can be no substitute for essential elements in life processes.

In agriculture the availability of oil (for traction) seems to be the factor where a resource scarcity could first appear.

The common elements: calcium, magnesium, sulphur, potassium, nitrogen

Some of the elements needed in agriculture (calcium, magnesium, sulphur and potassium) are so abundantly available that a global resource crisis is impossible, though production prices may eventually increase and regional needs in some cases may have to be filled through imports. Nitrogen as gas is also available in inexhaustible amounts, the limiting factor in this case is energy for converting nitrogen gas to usable forms.

Phosphorus

Phosphorus is no. 11 in the list of the most abundant elements in the earth's crust. Phosphate rock for fertilizer manufacture comes from a wide variety of deposits of phosphate minerals. Of these the sedimentary deposits (old animal bones) are the most important. Deposits in North Africa, with large reserves in Morocco, and those in Florida and North Carolina are all of sedimentary origin. There are also considerable sedimentary deposits in other parts of the world such as the Middle East.

Estimates of world reserves are about 14 thousand million tonnes of phosphate rock recoverable at current production costs, with another 20 thousand million tonnes recoverable at production costs up to 3 times current costs. In 1988, production was 153 million tonnes of phosphate rock. About 80 per cent of this was used for fertilizers, the rest for technical purposes.

Thus, known reserves and sub-economic resources should last for at least 200 years depending on future production rates. It is likely that further resources will be discovered. A real scarcity of phosphate rock seems very remote.

The micronutrients

Iron and manganese minerals are so abundant that a scarcity of these elements for agricultural purposes is quite unlikely.

Boron deposits are large but few and should last for more than 300 years; those of molybdenum for more than 150 years. Known resources of copper, selenium and zinc will last for some 30 - 60 years at the current mining rates.

The industrial use of all of these elements is overwhelmingly dominant. As scrap metal can be used for recovering micronutrients, they should be available for agriculture for a much longer period than the lifetime of known mineral resources.

References: US Bureau of Mines (1985).

4.13 Environmental and health impacts of fertilizer production

The local environmental issues

Fertilizer production, like all industrial activities impinges on the locality and its environment:

- directly through emission
- indirectly through heavy traffic, contribution to employment, etc.

The emissions from fertilizer production depend on a number of factors:

- the raw materials
- the processes
- effluent treatment
- standards of control, operation and maintenance
- factory history: old facilities were designed with less emphasis on pollution control than comparable new factories.

Only general indications of emission levels can be provided because of this complexity. Individual cases must be evaluated on their merits and the local conditions.

Emissions to the air and water

These will in most cases fall within the range given in Table 4.13.1. The upper part of the range can be found in old plants, while modern facilities show lower losses. Fertilizer factories also emit dust and industrial noise. As indicated, all compounds are not relevant for all installations, e.g., phosphate and fluoride come only from the production of phosphate fertilizers.

Table 4.13.1 EMISSIONS TO AIR AND WATER FROM FERTILIZER PRODUCTION

Emissions to air and water will usually be within the ranges:				
Air:	Ammonia 0 - 10 kg N/t N	Nitrogen oxides* 0.3 - 4 kg N/t N	Nitrous oxide 0 - 10 kg N/t N	Fluoride 0 - 0.12 kg F/t P
Water:	Ammonium and nitrate 0 - 5 kg N/t N		Phosphate 0 - 4 kg P/t P	Fluoride 0 - 1 kg F/t P

* $NO + NO_2$

Fertilizer factory, Sluiskil, the Netherlands.
Photo: Hydro Agri Sluiskil.

The general environmental issues associated with these pollutants are discussed in Sections 2.3, 4.3, and 4.6. Some 2-3 per cent of the world's fossil fuel consumption is used for fertilizer production; this generates a corresponding amount of carbon dioxide (Section 2.3).

As can be seen from Table 4.13.1, nutrient losses during fertilizer production are less than 1 per cent of the nutrients handled. The main environmental concerns associated with fertilizers are not at the production site but on the farm.

Solid waste

Gypsum (calcium sulphate) is a mineral which also occurs in nature. Some 10 tonnes gypsum/t phosphorus is formed as solid waste in phosphate fertilizer plants which use sulphuric acid for acidulation.

This gypsum also contains part of the other elements present in phosphate rock. It has been common practice in the industry to pump gypsum into the sea, where it gradually dissolves. Gypsum is a natural constituent of sea water. In some places, restrictions on the disposal of gypsum into the sea have been introduced. In other areas, disposal into the sea is permitted subject to restrictions on the cadmium content of the gypsum. Disposal on land is also used. Processes for cadmium removal in phosphate fertilizer production (Section 4.6) will eventually produce a waste that must be given appropriate attention.

Safety and occupational health

Production, transport and storage of ammonia and ammonium nitrate are subject to regulations designed to ensure safe operations and prevent fires and explosions (EC (1982)). Nitrate fertilizers can decompose in fire, e.g., in burning storage buildings, and release toxic nitrogen oxide gases. Storage of such fertilizers is subject to regulations. The risk of decomposition depends on the fertilizer composition. Some mixtures of fertilizer compounds can self-sustain decomposition on heating. The need for avoiding these compositions limits the range of possible formulations.

Dust exposure is the main occupational health problem in fertilizer manufacture. It is not an industry associated with unusual levels of cancer incidence. But use of asbestos for heat insulation was previously a common practice in industry. This was also the case in fertilizer production, and some tragic cases of lung cancer due to asbestos have occurred.

NPK fertilizer factory, Glomfjord, Norway.
Photo: K. Foss, Norsk Hydro.

4.14 Other and indirect environmental issues relating to fertilizer use

The present rural landscape with its flora and wildlife is in many areas the result of farming activities. Meadows often depend on grazing: if the land is not grazed, it can revert to scrubland and forests. When the productivity of meadows is increased through fertilization, fast-growing grasses will crowd out plants that thrive under less fertile conditions. Where this is done on an extensive scale, local varieties of plants and insects can be at risk of extinction.

Hedges around and between fields are important habitats for flowers, birds and insects. Mechanization of agriculture with the introduction of large machines has been associated with increased field sizes and the removal of hedgerows. This has happened on a large scale in parts of Germany and the United Kingdom.

Examples of other controversial topics in current agriculture are:

- specialized arable farming, perhaps with cereals as the principal crop
- a lack of genetic variety in crops
- depopulation of the countryside due to the mechanization of agriculture.

Availability of fertilizers is part of the foundation of current agriculture as described in Chapter 3 and fertilizer use is thus criticized also on these wider issues.

But good agricultural practices with extensive fertilizer use is not incompatible with an agriculture where more emphasis is put on maintaining suitable land for the preservation of local flora and wildlife.

As these matters for the most part are indirectly related to fertilizer use they will not be discussed in detail here, though they are important issues in the current debates on the future of agriculture.

Agriculture and forested land is a traditional combination in the landscape. From Telemark, Norway. Photo: K. Foss, Norsk Hydro.

CHAPTER 5

CURRENT FERTILIZER USE: ARE THERE ALTERNATIVES?

Environmental issues associated with current agriculture have intensified interest in new or different farming practices. Several directions are followed in this search:

- improvements of current practices designed to prevent nutrient losses and minimize the environmental impact of pesticides
- development of less intensive farming methods with reduced levels of fertilizer application and with the use of pesticides restricted to the control of severe infestations. The term "integrated agriculture" is increasingly used in Europe for such systems
- development of agricultural systems that reject the use of mineral fertilizers and pesticides as a matter of principle - "alternative agriculture" is a general term often used in Europe.

Efforts that are made within current agriculture to minimize environmental impact have been covered in preceding chapters. Here, we consider the background to, and performance of, the other agricultural systems and some of the associated issues.

5.1 Integrated agriculture

Integrated agriculture (Vereijken (1986, 1989), Zadoks (1989)) embraces several kinds of agricultural production systems that reduce the amount of purchased inputs, especially of fertilizer and pesticides. In the USA, such agricultural systems are called "Low Input Sustainable Agriculture", known by the acronym, LISA. (US National Research Council (1989), Potash and Phosphate Institute (1989), Hileman (1990)). This is a concept rather than a defined specific agricultural system. Several definitions are available, ranging from the very general to farming systems that use no purchased chemicals.

It should be emphasised here that sustainability is a major objective of all farming systems. Sustainable systems have four main characteristics:

- economic viability
- environmental soundness
- political and social acceptability
- conservation of natural resources.

Many systems of agriculture, including much of current agriculture, can claim that they follow this concept of sustainability.

Integrated farming systems usually have a crop rotation sequence that includes a legume to supply nitrogen and provide animal feed. They are generally less capital-intensive, smaller, and more diversified than is usual in current agriculture. The profitability of these systems will depend in part on regulations that may be imposed on the use of agricultural chemicals, in part on the ability of the farmer to operate a system with increased risks and a greater demand for labour. Some suggest a system of government payments in return for reducing the inputs of fertilizer and pesticide in order to reduce surplus production.

The environmental issues associated with these systems are the same as those in current agriculture. Their development is at an early stage at present. It remains to be seen if they confer substantial environmental advantages compared with properly conducted current farming practices.

5.2 Alternative agriculture

Different alternative agricultural systems exist. Various names are in use for such systems, e.g., natural, organic, biological, sustainable or ecological farming. We will generally refer to such systems as alternative agriculture, as is common European practice. To avoid confusion, it should be mentioned that the term is increasingly used in a broader sense in the USA, as a term covering both integrated agriculture and alternative agriculture in the more restricted European sense, Hileman (1990).

The alternative agricultural systems differ greatly in their basic ideas and recommended practices, but have in common that they reject the use of soluble mineral fertilizers and pesticides.

Less than 1 per cent of the farms in Western Europe are presently practising alternative farming. But in the last decade the movement towards alternative agriculture has gained in popularity and received official political recognition and support in the industrialized countries of the West. Today a main driving force is the market's demand for agricultural products made without the use of man made chemicals. Polls indicate that such products could take about 10 per cent of the market. The price premium for such products varies widely; 25 - 100 per cent is common.

The background and philosophy of alternative agriculture

A stimulus for this market trend is the current popular interest in alternative medicine and products made without inputs from the chemical industry. These topics have a long history.

The rapid development of natural sciences, of scientific agriculture, nutrition and health care, and the resulting social changes left some people dissatisfied for a variety of reasons.

They found scientific descriptions of nature too simplistic, and held that scientists split their subject into parts and study them in isolation, thus missing important interactions. A deeper understanding should be obtained by adopting a broader, holistic view where attention is focused on the whole and not on its parts, and where all possible mutual influences are taken into consideration. (Scientists regard this view of science as misinformed (Medawar and Medawar (1985))).

From such disagreements there developed in the first part of this century a variety of views different from those of science, of man and his interaction with nature, on the causes of disease, on the nutritional needs of man and animals, and on soil processes and their proper place in farming.

An example of such views is the opinion that substances produced by living organisms are in some respects different from the same compounds (e.g., nitrogen nutrients) produced by industry.

These widely different views and philosophies led to a variety of alternative agricultural systems: biodynamic, organic-biological, macrobiotic, Lemaire-Boucher, etc.

The other mainspring of the growing interest in and acceptance of alternative agriculture is the increasing environmental awareness amongst the general public.

Many people find aspects of current agricultural practices disquieting and objectionable: pesticide residues in soil, water and produce, increasing nitrate concentration in many ground and surface waters, landscape changes with reduced variety, animal husbandry methods that are perceived as unnatural, degrading and wasteful, depopulation of the countryside and costly production in excess of domestic needs in developed countries.

Objections to the use of fertilizers are not based solely on the perception of fertilizers as a cause of pollution, soil impoverishment and degradation, reduced plant resistance to diseases and diminished quality of the produce. It is also felt that the easy availability of fertilizers and pesticides has made possible practices that are regarded with distrust such as specialized farming and intensive agriculture.

Though there are many differences of principle among the various schools in alternative agriculture, there are also extensive and open collaborations and exchanges of views and experience. The International Federation of Organic Agriculture Movements (IFOAM) acts as a common forum. The various groups collaborate in marketing development, with standards and recognized labels for produce and rules for permitted practices. The EC is now proposing minimum standards and official inspection, control and labels for produce from alternative agriculture. Some insight into the practices and philosophies of alternative agriculture is useful when one is evaluating the arguments, observations and claims for these systems.

Organic agriculture

This is now a rapidly expanding sector. As a movement it developed gradually from views expressed in the first part of this century, and even earlier (Conford (1988)). Its adherents are concerned about an agriculture dependent on non-renewable resources and about "unwholesome" food with residues of chemicals. They fear degradation of the soil and are dismayed at many of the aspects of animal treatment in intensive husbandry.

Organic and other alternative agricultural farms are of various types depending on local conditions. Some are located in areas not readily suited to arable crops, have most of their land as grass-clover meadows and produce mainly milk and some meat. Some small farms are specialized vegetable producers. But more typically, an alternative farm will practice mixed animal-arable farming with some 40 per cent of the land kept as grass-clover leys. Some of this is permanent grass; the rest is ploughed every 2 - 3 years as part of the rotation. The arable part is used for cereals, root and fodder crops, legumes and field vegetables in rotation. Part or all of the cereal, roots and legumes may be used as feed for the animals. The stocking rate should match the feed produced on the farm, e.g., in Sweden about 0.6 cows or equivalent per ha used for feed production. All animals have access to grazing, none are permanently confined to stables.

Soluble mineral fertilizers are not allowed, especially not nitrogen. Rock phosphate and other natural minerals with a low solubility can be used. Weeds are removed or damaged by mechanical soil treatment or the use of fire. Extensive crop rotation and intercropping are adopted while monocultures are avoided. For a description of practices on a large organic farm in the UK, see Wookey (1987). The views of the movement on environmental effects of farming systems were reviewed by Arden-Clarke and Hodges (1987, 1988).

The organic agricultural movement comprises groups that differ in their views on inputs of manure from other farms. Some restrict such inputs to manure from farms also practising alternative agriculture. Others permit substantial purchases of animal manure from current farming enterprises provided these follow recognized standards for animal welfare. The latter group support their own production by fertilizer use on these other farms.

Some farmers are allocating fields where they temporarily produce crops for one or a few years without using fertilizers and pesticides. They then market the produce at a premium. However, the organic agricultural movement does not recognize such produce as organic.

Alternative agricultural systems based on views different from those of science

As some of these systems are old and have a traditional standing we include a short description of their philosophies and views.

Biodynamic agriculture

This must be seen in the context of anthroposophic beliefs. The basic formulation of these beliefs, and of their application to agriculture, was made by Rudolf Steiner inspired by the ideas of Goethe. He taught that life has both material and nonmaterial components. The latter are due to cosmic forces from the stars and planets, the astral forces. These impose rhythms and shape on life and nature. Other aspects of the cosmic forces are the spiritual and the life or ethereal forces. The latter work especially through plants. Tillage creates a chaos that prepares the field for the influence of external enlivening and shaping forces. Sowing should preferably be done according to when the moon passes through the proper signs of the Zodiac (e.g., the signs: Libra, Aquarius and Gemini for flowering crops), so that the crops benefit from the sidereal rhythm.

Biodynamic preparations are used in minute amounts to stimulate plants, enliven the soil and mediate the external forces. One of these is made from cow dung that has been completely composted. Some compost is placed inside a cow horn and buried in the soil during the winter. Steiner claims that the horn can collect, conserve and concentrate ethereal and astral forces. The horn therefore ensures that the humus it contains is enriched through the shaping forces of the soil. Before it can be used the substance is stirred rhythmically for an hour in water, usually about 100 g in 30 - 40 litres of water. The rhythmic stirring is said to activate the preparation and transform the forces from a resting to an active form. The preparation must be used within a couple of hours after the procedure is finished in order to avoid wasting the forces. About 100 g per ha are applied to the soil surface in early spring afternoons with 2 - 4 applications during spring and summer. It is said to stimulate germination and growth processes.

Organic fertilizers are used, not for the direct benefit of the plants, but to intensify life processes in the soil. Soluble mineral fertilizers are rejected because they are thought to reduce interactions between soil and plant.

The movement is strongest in Germany and Switzerland, but also has followers in other countries, e.g., Norway and Sweden.

Organic-biological agriculture

This is based on the views of Hans Peter Rusch from Germany and Hans Peter Müller from Switzerland on the origin of disease and health and on what constitutes a healthy diet. They claim to have identified specific strains of bacteria that are beneficial to man and to plants. These "physiological bacteria" are said to be carriers of health-giving "macromolecular living substances". They teach that it is important that food is rich in these elements, and that they are taken up by the plants from the soil. The organic-biological agriculture is designed to facilitate plant uptake of such elements, and their return from man and animals to the plants through composting and soil treatment methods.

Fresh organic material is considered poisonous to plant roots. Organic material must therefore only be added to the surface, and the soil worked only to a very shallow depth so that fresh plant material does not come in contact with roots. Rock phosphate, lime and finely ground stone are permitted.

Organic-biological agriculture is mainly practiced in Switzerland and Germany, but can also be found in several other countries.

Other systems

Other similar schools of thought in alternative agriculture exist, e.g.,

- Lemaire-Boucher agriculture
- macrobiotic agriculture.

Boeringa (1980) gives a more detailed description of these.

References: Boeringa (1980), CAST (1980).

5.3 Main problems in alternative agriculture

Nutrient supply

The soil processes governing nutrient uptake, transformation and losses are the same whether a farm follows current or alternative practices. These processes are described in the preceeding chapter.

The difference is in the way nutrient losses are compensated. In alternative agriculture, losses are compensated through:

- growing legumes for their nitrogen fixation
- application of ground mineral rocks (e.g., stone, phosphate rock, limestone) to supply phosphorus, potassium and other elements.

The main problems with nutrient inputs in alternative agriculture are:

- nitrogen is generally in short supply
- soil reserves of phosphorus and potassium can be depleted to levels that reduce yields.

As a matter of principle most systems of alternative agriculture reject all use of soluble mineral fertilizers, including those that are isolated from natural brines or minerals, e.g., potassium chloride and Chile salpeter (sodium nitrate).

Soluble fertilizers are regarded as detrimental to soil life and proper crop development, as they give "unnatural" soil conditions through enhanced nutrient concentrations. This is said to disturb the soil's processes and ecology and to give an unbalanced uptake of nutrients by the plants. Soil nutrient inputs should instead enhance soil nutrient reserves; the farmer should "feed the soil and not the plant". The supply of plant nutrients should derive from mineralization as a natural process.

Experimental support for this view is lacking. By rejecting the use of fertilizers, alternative farmers are restricting their ability to tailor the supply of plant nutrient to a crop's requirements.

Nitrogen

Nitrogen is often a principal yield-limiting factor in alternative agriculture (Kahnt (1986)). The biological fixation of nitrogen through the extensive growing of legumes is the nitrogen source. This topic is described in Chapter 1. Part of the supply of nitrogen to the legume comes from nitrogen fixation, while part derives from the uptake of mineral nitrogen from the soil. The relative proportion varies according to species, varieties and soil conditions. Table 5.3.1 gives a list of the ranges of published amounts of nitrogen fixed for field-grown legumes.

Table 5.3.1 AMOUNTS OF NITROGEN FIXED FOR SOME FIELD-GROWN LEGUMES AND THE AMOUNTS OF PLANT NITROGEN DERIVED FROM THIS FIXATION

Species of legume	N fixed kg/ha Range	Per cent plant N from fixation Range
Soybean (*Glycine max*)	10 - 115	23 - 67
Broad bean (*Vicia faba*)	138 - 237	64 - 91
Common bean (*Phaseobris vulgaris*)	25 - 125	37 - 68
Lentils (*Lens culinaries*)	129 - 192	72 - 86
Peas (*Pisum sativum*)	83 - 189	62 - 82
Clover (*Trifolium sp.*)	180 - 200	85 - 100

Source: Rennie (1985).

Even higher rates of nitrogen fixation have been recorded, e.g., up to 600 kg N/ha for lucerne (alfalfa).

The pods of legumes are rich in nitrogen as protein. If they are harvested for sale, the value of the crop as a nitrogen source for the succeeding crop is diminished. The nitrogen in legumes fed to animals is partly recycled through manure; this provides a nitrogen supply based on legumes suitable for application. Hence farm animals have a central and necessary role in alternative agriculture. But animal husbandry and manure collection, storage and use are also activities associated with major losses of nitrogen.

In a permanent meadow with clover, nitrogen from legumes becomes available to the grass when the legumes die, and the permanent root system ensures that the nitrogen which is liberated is not wasted.

Attempts are made to extend mixed legume systems to arable crops by planting cereals together with legumes and grass (as a catch crop). However, living legumes contribute little nitrogen to their neighbours in the same season. Plant breeders are currently attempting to develop "leaky" legumes for use in mixed cultures. But cereals take up most of their nitrogen during the first 8 - 10 weeks after germination, while legumes need a well developed plant for extensive nitrogen fixation. Thus the two crops are not well matched. The need for light, nutrients and water both for crop, clover and grass makes mixed cultures somewhat less productive than monocultures.

Growing legumes for their nitrogen input requires space and time in the rotation, and thus reduces the total productivity of the farm. The variable and unpredictable productivity of grass-legume swards (van der Meer and Baan Hofman (1989)) is also a problem.

Legumes need much phosphate and potassium for proper development. When slow but inevitable depletion of soil reserves occurs, the basis for biological nitrogen fixation is eroded. Attention to the soil condition with appropriate liming is required as legumes contribute to soil acidification, which is detrimental to biological nitrogen fixation.

In alternative agriculture, emphasis is placed on the return of organic waste to the soil. It is common to compost crop waste and manure together. Composting is an aerobic microbial process where the easily degradable organic substances are digested and converted to stable, slowly degradable organic matter. The process generates heat and thus kills most weed seeds and pathogens. The bulk of the material is reduced but some nitrogen is lost to the atmosphere as ammonia. Losses of up to 40 per cent of total nitrogen are reported.

Urea from urine in liquid parts of manure is soluble and rapidly available to plants. But most nitrogen in alternative agriculture is present in organic matter that requires mineralization, and some of the input will carry over into the next seasons as described in Section 4.3. It is difficult to estimate the nitrogen input to a crop in alternative agriculture. In contrast, nitrogen in fertilizers can be applied in amounts which more readily match crop

demand. Such differences make it difficult to conduct matched experiments where equal nutrient amounts are supplied in alternative and current systems.
Reference: Heichel and Barnes (1984).

Clover under-sown in wheat, providing some nitrogen for the next year's crop. Photo: T. A. Breland, Department of Biotechnological Sciences, Agricultural University of Norway.

Nitrate leaching

Some features of alternative agriculture are conducive to low levels of nitrate leaching:

- a substantial part of the land is in grass swards
- the amount of nitrogen generated on the farm and applied as manure is so limited that overall application rates are low.

On the other hand, some practices that are basic to alternative agriculture:

- ploughing grass-clover leys
- legume growing
- manure use
- fallowing of land for weed control

all imply significant risks for nitrate leaching.

Rushall farm is a well-known alternative farm in the UK (Wookey (1987)). The principal nitrogen source is grass-clover leys that form part of the rotation. Measurements indicate that the ploughing of these gave nitrate losses of 99 kg nitrogen/ha • year, compared to the over-all average annual nitrate losses of 19.7 kg nitrogen/ha. This average loss is broadly similar to what is expected from a mixed farm using current agricultural practices (Davis and Barraclough (1990)).

Danish studies also indicate that alternative agriculture does not necessarily offer less nitrate leaching than current practices (Østergaard (1988)). In both systems measures must be taken to prevent leaching by adhering to good agronomic practices. Such practices appear more important than the choice of agronomic system.

Phosphate

Alternative agriculture does not permit the use of chemically treated rock phosphate and states that ground phosphate rock is sufficiently reactive for use as a nutrient source.

Plants can only take up phosphate from the soil when the phosphate is dissolved in water in the soil. In order to supply the immediate phosphate needs of plants in a soil with insufficient phosphate, soluble phosphates must be made available. Ground rock phosphate is generally not sufficiently reactive for this purpose, though certain rock deposits have sufficient reactivity to give immediate plant phosphate response on acid soils such as occurs in some forests and in the tropics. Tunisian (Gafsa) and North Carolinian rock can be used on soil where the pH is below 6.5. Only a few per cent of traded rock is of this quality. Such phosphates cannot satisfy world needs for soluble phosphate, though they can supply the present alternative farms.

Plants can react to phosphate deficiencies by making the soil near their roots more acidic; this can increase the solubility of phosphate minerals. Plants vary in their ability to make use of phosphate rock as a phosphorus source; clover benefits more than cereals. In spite of this, the efficiency of most ground phosphate rock as a plant nutrient source is very low; results with cereals are notably disappointing (Khasawneh and Doll (1978)).

Much of Europe's arable land has now reached a level of phosphate reserves in the soil where only maintenance dressings of phosphate can be adequate for a long period. The efficiency of ground rock phosphate for maintenance of phosphate status is thus a matter of some dispute. The efficiency will vary depending on soil conditions, (particularly calcium content and pH), types of plants, type of rock and rock treatment (e.g., fineness of grinding). However, grinding alone is not sufficient to induce reactivity.

Cadmium and other unwanted elements in phosphate rock are discussed in Section 4.6. The cadmium in ground rock phosphate cannot be removed by mechanical means. Phosphate rock resources have been covered in Section 4.12. Both current and alternative agriculture depend on these external non-renewable resources for the maintenance of the soil's fertility.

Basic slag from steelmaking contains phosphate. This phosphate (e.g., in Thomas phosphate) is just as available to plants as phosphate in fertilizers. Thomas phosphate is approved for use in alternative agriculture. It also increases soil pH and calcium status. However, the Thomas process is no longer used in Western Europe and basic slag currently supplies only 0.3 per cent of world phosphate demand. Thus, the use of such materials cannot supply the need of world or local agriculture for fertilizer phosphates.

The soil reserves of phosphate and potassium can be extended by growing green manure crops with especially deep roots (e.g., lucerne (alfalfa)) that can tap the subsoil. This can postpone mineral depletion of the soil but not eliminate it.

As discussed in Section 4.2, the roots of most crop plants grow in association with fungi. These mycorrhizal associations extend the volume of soil that the plant roots can explore for uptake of nutrients, notably phosphate and potassium. Mycorrhiza are especially important in alternative agriculture. They enable plants to make efficient use of available phosphate in the soil, but cannot compensate for gradual depletion of the nutrient.

Other nutrients: potassium, magnesium, etc.

Soil minerals originate from rocks by grinding (e.g., action of ice and water) and slow weathering. Alternative agriculture advocates copying nature when additional mineral supply is needed for proper plant growth, and recommends the use of ground rock as the source of potassium and trace elements. The ability of ground rocks to dissolve and contribute to the supply of soil nutrients varies with the mineral, but is generally very low even in acid soil water (pH below 6) (Blum et al. (1989)).

Hög (1985b) reports that in the most favourable cases, 1t/ha was required to provide 20 kg potassium/ha.

Thermal weeding: emerging weeds are damaged, giving the crop of onions a competitive advantage. Propane gas is used as fuel.
Photo: The Department of Agricultural Engineering, Swedish University of Agricultural Sciences.

Weeds, insect infestations and plant diseases

Weeds, insects and plant diseases cause major problems both in current and alternative agriculture. The first line of defence is keeping the intensity of the infection to a minimum. Farms that are negligent in this respect represent potential centres of infection for the surrounding fields. On the other hand such a farm benefits from its neighbours' efforts to keep pest infections at low levels.

Alternative farmers do not strive to keep their crops free from weeds but to control the infestation by non-chemical means. They may also make use of some weeds, e.g., for tea, salads, medicine and extracts (e.g, nettle) reputed to have effects against pests.

The following indirect and direct methods are used for pest control in alternative agriculture:

some of these methods are also standard practice in current agriculture, e.g.,

- ploughing, harrowing and other mechanical measures of cultivation
- adjustment of soil conditions through liming
- use of healthy and pure seeds
- use of disease resistant cultivars

some measures are now practised more extensively in alternative agriculture than in current agriculture especially:

- crop rotation
- surface covering with crop residues
- mixed cropping in order to give the next crop a competitive advantage over weeds after harvest of the first (usually cereal) crop

certain measures are now almost exclusively practised in alternative agriculture:

- use of crops with an efficient canopy closure (e.g., potato) in order to shadow out weeds
- damage to weeds through burning with a gas flame ("thermal weeding")
- delayed sowing
- periodically keeping the land fallow for part of the growing season.

Manure usually contains weed seeds. Their number can be greatly reduced by composting; thus composting is a recommended sanitary process on alternative farms. But it also has the disadvantage of nitrogen loss.

Generally, these methods are more cumbersome and less efficient than the use of agrochemicals and require a wider range of equipment. But they also have merits as elements in integrated pest control systems and some may become more generally used in future agriculture.

Biological weed killers are an alluring possibility for the future, but are currently available only for a few special weeds.

Alternative agriculture lacks efficient methods for emergency treatment of attacks on crops by insect or plant disease. The crops are thus at greater risk than in current agriculture. The risk varies depending on the crop, but is especially severe in fruit crops.

The efficiency of some of the measures such as plant extracts and biodynamic preparations is contentious.

Direct control of insects and fungi by predators (e.g., ladybirds) is slow and erratic and is most suited for gardens and green-houses and for a limited number of insects. It is difficult to ensure an adequate population of the predators in a given location.

Mixed planting or sowing is recommended for plant protection on alternative farms. One example is carrots together with onions. While some reports indicate reduced attack from the carrot fly, others have not been able to confirm these findings.

Some of the varieties (e.g., of potatoes) commonly grown in current agriculture are too susceptible to disease to be used when agrochemical protection is renounced. The alternative farmer is therefore somewhat restricted in the selection of cultivars.

As discussed in Sections 4.2 and 4.7, alternative agricultural farms tend to have less problems with erosion, more earthworms and better conservation of the soil's organic matter than in specialized arable farming. The inclusion of grass leys in the rotation and practice of mixed animal-arable agriculture is the principal cause.

4 Productivity and economics in alternative farming

Cropping patterns

Cropping patterns in alternative agriculture differ from those in specialized current agriculture. Grass-clover leys and fodder crops are necessary and extensive rotations are more common. Farms in alternative agriculture tend to have markedly less grain and meat production than those in current agriculture, and produce a relatively larger proportion of dairy products, potatoes, pulses and some vegetables.

Danish data illustrate the difference in cropping patterns between current and alternative agriculture, Table 5.4.1.

Table 5.4.1 CROPS ON CURRENT AND ALTERNATIVE FARMS IN DENMARK, 1981

	Per cent of area	
Crop	Current	Alternative
Cereals	62	33
Herbage	14	36
Vegetable	1	8
Potato		5

Source: Hög (1985b).

Part of the difference in emphasis on vegetables probably relates to differences in farm size, since 43 per cent of the alternative farms had less than 5 ha of cultivated land. The comparable figure for current farming was 13 per cent. A Norwegian study showed similar proportions (NLVF (1983)).

Productivity

Productivity is very much a matter of skill and committment to a particular system. Comparisons between alternative and current farming systems are therefore very difficult to make. The product range and ratios are very different. Yields of the same crop from similar fields under different management can be compared, but such comparisons cannot be extended to estimates of total farm productivity when part of the land is used mainly for ensuring nutrient supply. Further, yields obtained in alternative agriculture can reflect high soil fertility from nutrient reserves due to natural abundance or previous fertilization. Such yields may not be sustainable. Generally, it is to be expected that low input agriculture cannot sustain a high output.

As discussed in Section 2.3, the world-wide average cereal yield in 1950 was about 1.2 t/ha. A combination of fertilizers, plant protection, irrigation and new varieties have increased this to about 2.5 t/ha. Renouncement of fertilizers and pesticides should return the yields to a level more like the previous than the present one.

Nutrient supply on an alternative farm has to rely on the farm's own resources with nitrogen fixation by legumes and with some import of rock phosphate and finely ground minerals of uncertain efficiency. Any purchase of animal feed, manure or wastes from other farms represents an external input of nutrients and should be taken into account when studying the nutrient balance on an alternative farm. The problem is to estimate the export potential from such a farm when it is in nutrient equilibrium.

Gesslein and Jansson (1987) report from an experiment in Sweden on a soil unfertilized since 1942 but cultivated with manure and clovers as generally practiced in alternative agriculture. The yields of most crops were 20 to 30 per cent of the yields of current systems. Adequate liming and some fertilization with phosphorus and potassium raised the yield somewhat when only legumes and manure were used as the nitrogen source.

Wookey (1987) estimates from his own practical experience on a large organic farm in the UK that the output is half of that in current agriculture. A Swedish study (described in Section 4.2, yields from natural fertility) supports this estimate.

A study in the Netherlands (Boeringa (1980)) suggested that production could gradually drop to 20 per cent of the production level in 1980 if no mineral fertilizers are used, all manure is used on the farm that produced it, and no import of concentrates takes place. The estimate applies to nutrient-poor sandy soils. Some 40 per cent of cultivated soils in the Netherlands are of this type. If soil phosphorus and potassium levels are kept adequate, a production of about 65 per cent of current production could be maintained with dairy items as a main product.

If the soil has large nutrient reserves the rate of nutrient depletion may be slow, and higher yields than this can be obtained as long as the reserves last.

Root crops and vegetables are important crops in alternative farming, since they provide variety in the rotation. But they are highly susceptible to insects and diseases. Hög (1985b) reported that the marketable yield of potatoes could be less than 50 per cent of normal, and for some vegetables, e.g., cabbage, even lower. These crops thus carry considerable risks of heavy losses.

Experience gained in the development of alternative agricultural practices may be of value for the improvement of traditional agriculture in developing nations. But with the need for rapid increases in agricultural production in these nations, the prospects for satisfying these needs through adoption of alternative practices appear doubtful.

Organic gardening appears more attractive and practicable than alternative farming. This is because:

- inputs of organic matter in sufficient amounts can easily be obtained from outside or from parts of the property not used for gardening
- manual weed control is feasible
- resistant varieties can be chosen independent of market preference
- severe pest infestations and poor yields do not imply food supply difficulties or financial problems.

Labour

Vereijken (1986) reported on the differences in the Netherlands between current, "integrated" and biodynamic farming. Although the yields were greatly reduced in biodynamic farming compared to the two other systems, gross incomes were 20 per cent higher due to higher produce prices. However, labour costs were about three times as high resulting in a large net loss. Böckenhoff et al. (1986) found that German alternative farms had an increased labour need of about 35 - 40 per cent compared to current farming. In contrast, Wookey (1987) reports no increased labour needs for the Rushall organic farm.

A Danish survey (Hög (1985b)) gives conclusions from investigations in various countries. The results vary but all conclude that the labour requirement per produced unit is greater in alternative than in current farming.

Economy and marketing

Farmers practicing alternative agriculture have a greatly reduced outlay for fertilizers and agrochemicals compared to current farming enterprises. But as this outlay is only a part (some 25 - 50 per cent) of total farming expenditure the cost reduction does not compensate for reduced farm output, increased risk of crop losses, increased labour needs and costs for the greater variety of machinery in mixed farming.

There are conflicting reports on the profitability of alternative agriculture. Some report serious losses from organic biological farming in Sweden (Gesslein and Jansson (1987)) and the Netherlands (Vereijken (1986)) in spite of premium prices for the produce. Others report that alternative farming shows economic results about equal to those of current farming provided that a substantial premium is obtained for the product (Jaep (1986), Wookey (1987)).

German experience indicates that successful alternative farms should preferably be located on good agricultural land near ready markets where premium prices can be obtained (Jaep (1986)).

The premium generally ranges from 25 - 100 per cent. In Sweden, prices to consumers for alternative produce in the autumn 1989 were about 50 per cent higher than prices for comparable products from current agriculture (Anon (1989b)). This premium market is under active development with officially recognized labelling requirements. It is mostly cereals, vegetables, fruit and eggs that can command premium prices at present, but meat, milk and dairy products are entering this market to an increasing extent.

Today, demand exceeds production - less than 1 per cent of European farms follow alternative practices. The economy of alternative farming on a large scale is critically dependent on sufficient marketplace acceptance of food with a substantial price premium, or on public subsidies through the tax system.

5.5 Nutritional quality, wholesomeness and taste of produce from current and alternative agriculture

As described in Section 4.10, quality of agricultural produce refers both to visual outer quality and to the chemical constituents (inner quality). Evaluation of alternative produce principally concerns the latter, especially nutritional value, absence of noxious compounds and taste. We will here focus on comparisons between produce from current and alternative agriculture for those parameters of quality which can be measured using validated analytical techniques.

Some alternative movements also include in the quality concept items that cannot be measured by generally recognized and validated methods, such as life forces and inner harmony in the produce. Ethical considerations, e.g., animal welfare, are also sometimes used indirectly as a quality standard, for example in free-range eggs.

Produce content of vitamins, minerals and other essential dietary factors

It is well known that the chemical composition of plants is influenced by growing conditions, including nutrient supply and balance (Section 4.10, Hög (1985a), Mengel (1979)); but the existence of consistent differences in the content of essential dietary factors between produce from current and alternative agriculture is a matter of dispute.

Shupan (1974) reported from matched plot experiments that roots and vegetables given biodynamic compost had more dry matter, sugars and vitamin C, methionine (an essential amino acid), potassium and calcium than produce fertilized with mineral fertilizers. The latter produce had a higher content of free amino acids, sodium and provitamin A (carotene).

However, such effects are connected with extremes in nitrogen supply. With practical levels of fertilizer application, quality effects are seldom appreciable (Greenwood et al. (1980), Geissler and Geyer (1980)). Nilson (1979) found that the small differences seen depended on the amounts of nutrient given and not on the type of nutrient supply - mineral or organic. Hansen (1981) and six other reports listed by Nilson concur.

These authors conclude that provided

- the products are at the same stage of development and maturity at harvest, and
- no deficiency or excess of any essential plant nutrient has been present

no differences between vegetables grown by current and alternative practices can be expected.

Nitrate in vegetables

The topic of human nitrate intake was discussed in Chapter 4.3. Usually most of it comes with root and leaf vegetables. Cereal grain contains practically no nitrate. Vegetables vary greatly in their nitrate content. The principal influences are:

- species
- part of the plant used
- amount of sunlight during the last phase of crop growth. Winter grown leafy vegetables tend to be high in nitrate.

Vegetables therefore show a wide range of nitrate content which can vary by a factor of 10 or more:

Table 5.5.1 EXAMPLES OF RANGE OF NITRATE CONTENT IN VEGETABLES

Vegetable	mg nitrate/kg fresh weight
Beetroot	497 - 6631
Lettuce	90 - 3520
Cauliflower	40 - 740
Cucumber	40 - 445
Potato	7 - 360

For a comprehensive survey with extensive details, see Corré and Breimer (1979).

The nitrate content also depends on the nitrogen supply. Reduction of the nitrate content of leafy vegetables through reduced nitrogen supply markedly reduces yield (Greenwood and Hunt (1986)). Excessive application of nitrate can notably increase the nitrate content of leaf and root.

There is some disagreement on the effect of organic nitrogen supplies on vegetable nitrate content. Vogtman et al. (1984) found that composted manure gave a lower nitrate content in produce than fertilizers, while Gysi et al. (1985) found that lettuce had an enhanced nitrate content when grown on soil with high levels of organic nitrogen.

In actual practice there does not seem to be much difference between the nitrate content in produce from the two production systems, an observation supported by investigations by Hansen (1981) and Diaper and Lawrenson (1986). A recent paper from scientists at UK/German research institutes for alternative agriculture (Stopes et al. (1988)) found that lettuce from current and alternative growers had about the same nitrate content, though the probability of finding peak nitrate concentration values may be lower with alternative produce. There are marked varietal differences in nitrate accumulation for some vegetables, e.g., lettuce. Thus plant breeding may provide a way for reducing the nitrate content.

Taste and aroma

There is a wide-spread impression that produce from alternative farms has a better and more natural taste than the produce of current agriculture. However, blind testing in Denmark and Germany with a variety of vegetables showed no consistent preferences for the current or alternative products (Hansen (1981), Vetter et al. (1983)). Other similar studies (listed by Hög (1985b)) are generally in accord with these results.

Only a few comparisons of milk from alternative and current farming systems are available. Knoppler and Averdunk (1986) found no significant quality differences, while Boeringa and de Vries (1989) reported that the panel tended to prefer milk from the farm practising current agriculture.

Pesticide residues and other unwanted food components

Since pesticides cannot be used in alternative agriculture, it is to be expected that levels of pesticide residue will be low, and originate mainly from general aerial deposition with rain and dust.

Pesticides are evaluated by government institutions before their use is permitted. For a description of the data required by European authorities before a pesticide can be evaluated, see Council of Europe (1984).

Regulations for the use of pesticides in current agriculture take into account the need for ensuring low residual levels in produce. Thus residual levels tend to be low also in current agriculture. Instances of pesticide residues in excess of permitted levels do occur and are a cause for concern, but they are exceptions, not the rule. Danish results can be taken as an example. The authorities found for the period 1983-87 that 0.5 per cent of Danish produced vegetables and 1.6 per cent of the imported produce had pesticide residue levels exceeding Danish permitted limits. In most cases, residue excesses are said to be minor. Pesticide residues were not found in the large majority of the samples. Where residues were found, the average level was much below the permitted limits (Danish National Food Agency (1989)).

In accordance with this, the average dietary intake of pesticide residues in UK diets was well below 1 per cent of acceptable daily intakes for all pesticides reported except for dieldrin where the average daily intake in 1984-85 was about 7 per cent of the acceptable level (MAFF (1989)). Dieldrin is no longer used and the environmental levels should gradually decrease.

This gives some perspective on the issue of pesticide residues. Nevertheless, the risk of having residues in amounts exceeding statutory limits is necessarily greater in current than in alternative produce.

Fungal toxins in food represent a potential health problem in the tropics. Such toxins from mouldy foods are possibly a cause of liver cancer in some areas. However, in the developed countries both current and alternative agricultural products usually show undetectable levels of these toxins; there seems to be no difference between the two types of produce in cases of mould infestations in those few instances where these compounds are found. This is in accordance with what is known about fungal growth and toxin production: the dominant factor will be storage conditions for the produce.

The subject of cadmium was discussed in Section 4.6. The cadmium content of farm produce depends mainly on soil acidity. Consistent differences in cadmium content between current and alternative produce have not been found.

Technical quality criteria

One quality difference between current and alternative agriculture is well recognized: wheat from alternative farms tends to have less protein than wheat produced by current agricultural practices. This is to be expected in view of the limited nitrogen supply on alternative farms, as also seen in the lower yields. Alternative farmers can have difficulties in meeting the EC protein criteria for premium breadmaking wheat, though Wookey (1987) reports good baking experience with wheat that did not meet these criteria. The EC protein criteria for wheat have been set to ensure good baking properties. The differences seen have no nutritional significance in the context of the Western diet where protein intake generally exceeds minimum needs.

Another quality difference often mentioned is that alternative produce has a lower water content and better storage properties than current produce. Conflicting results are reported, thus Nilson (1979) and Hansen (1981) did not find such differences while Samaras (1977) found less loss and microbial degradation in alternative produce.

Unbalanced fertilization can be detrimental to quality properties including storage ability but balanced nutrient supply is regarded as necessary to ensure good storage properties. Other important factors influencing quality (Henze and Hansen (1988)) are:

- soil type
- weather
- ripeness and conditions at harvest time
- handling and conditioning (e.g., drying) after harvest
- conditions during storage, transport and marketing.

The quality of the produce as experienced by the final customer and consumer thus depends on many factors. Minor differences in produce quality due to differences in nutrient supply will be submerged in the effects produced by the many other, more important influences on produce quality.

Comparisons of marketed current and alternative produce in Germany (Vetter et al. (1983)), Denmark (Jensen and Leth (1985)) and in New York, USA (Gourdine et al. (1983)) found essentially no quality differences. These studies have been criticized because some of the products sold as alternative might have been produced by current methods and fraudulently labelled, thus distorting the results. However, other, though smaller, studies where care was taken to control this factor have given the same results (e.g., Bulling (1986/87)).

It would be valuable for the guidance of customers if similar large-scale studies could be made when strict labelling and control arrangements for alternative produce have been put into general use. However, in view of what is already known about the topic, notable and consistent differences are not expected.

Alternative agriculture and health

Animal health

Methaemoglobinaemia and grass tetany from incorrectly fertilized herbage were discussed in Sections 4.3 and 4.5. Current interest in possible connections between the mineral content of feed and animal health was also referred to in Section 4.5.

Statements have been made that animals on alternative farms or given feed from such farms have fewer fertility problems, remain productive longer and in general have fewer health problems than animals in current agriculture.

There are only a few mostly old and conflicting studies and they invite a variety of explanations. Animal husbandry conditions in current and alternative agriculture differ in more aspects than feed production. The issue thus remains open (Augstburger et al. (1989)).

Food and human health

It has long been known that food and eating habits are related to health and disease. A desire for healthy food is part of the market acceptance of alternative produce.

But van Stavern et al. (1985) compared the size and development of Dutch preschool children from homes using food from current agriculture with children from homes where the food was vegetarian, biodynamic or macrobiotic. They found no indication of superiority in the latter practices. The children from the macrobiotic homes were undernourished.

More attention to diet and lifestyle can contribute significantly to improved health in the population. Table 5.5.2 indicates important dietary factors implicated in human health and diseases.

Table 5.5.2 IMPORTANT DIETARY FACTORS IMPLICATED IN HEALTH AND DISEASE

Disease	Risk factor	Protective action
Malnutrition	- monotonous diet (from poverty or ignorance)	- varied diet
Cardiovascular disease	- overeating	- reduce sugar and fat intake - ample fish
Cancer	- excessive fat, salted, smoked foods - alcohol	- liberal intake of fruits and vegetables
Allergies (disposed individuals)	- some common foods (e.g., milk, egg) - some spices and other additives	- careful labelling of processed food

The list could be expanded, e.g., food and water borne infections remain a serious problem. A description of diseases related to nutrition in Europe was recently published by WHO (James et al. (1988)).

Knowledge about diet and health are reflected in dietary recommendations. There is no standard European diet. Eating habits vary and depend on national traditions and dietary recommendations vary somewhat between nations. But the current recommendations from the Norwegian National Nutrition Council (1989) are typical of present advice and knowledge about how a European diet should be made more healthy.

The Nutrition Council recommended the following dietary changes:

- less fat, primarily less saturated fat. Fat should not supply more than 30 per cent of the energy in the diet
- more starch and less sugar. Carbohydrates should altogether supply 55 - 60 per cent of the energy in the diet, whereas sugar should not supply more than 10 per cent
- more fibre. The intake of dietary fibre should be at least 3 g per MJ, which for adults correspond to 25 - 30 g per day
- less alcohol.

In order to achieve these changes and improve their diet people are advised to consume:

- more cereal products, especially unrefined, wholemeal products
- more fish
- more potatoes
- more fruit and vegetables
- leaner meat and low-fat milk products at the expense of higher fat products
- less margarine and butter
- less sugar and sugar products (sweets, confectionery)
- less salty food, and less use of salt in food preparation and as a condiment
- less alcohol.

These are the real and substantiated issues concerning food and health in Western society. The present problems derive from composition of the diet and amounts consumed, not from the way food is produced on the farms.

FINAL COMMENTS

Agricultural practices, including fertilizer use, raise a wide variety of environmental concerns that deserve serious consideration. But all forms of agriculture modify the environment. The challenge is to find the proper balance between costs and benefits to society and to nature.

We conclude from our analysis of the issues that:

- Fertilizers are required in order to maintain soil productivity and keep agriculture sustainable

- Global food production must be increased in order to feed an increasing world population. For this, appropriate use of water and fertilizers and measures ensuring crop protection are indispensable

- Intensive use of existing arable land is a prerequisite for maintaining large areas of potential agricultural land in their undisturbed natural state

- Current agricultural usages have some undesirable environmental consequences, e.g., nitrate leaching to waters, soil erosion, eutrophication and other issues described in this book. Good agricultural practices can minimize but not eliminate such effects. Proper handling and recyling of manure is very important

- Correct use of fertilizers is part of good agricultural practices, but untimely, unbalanced or excessive use is wasteful and can cause environmental problems

- Good agricultural practices are not static concepts. Farming practices are always evolving, with science providing guidance and understanding of complex relations. Increasing emphasis on environmental issues is part of this development

- Alternative agricultural systems produce less and more costly food than current agriculture and the benefits to consumers, farmers and the environment are questionable.

Farms form part of our environment, and the relationship between agriculture and the rest of nature is highly complex. Careful studies with proven scientific methods must form the basis for future actions and recommendations.

REFERENCES:

Adriano D.C. (1986) Trace elements in the terrestrial environment. Springer Verlag, Heidelberg.

Agrios G. N. (1988) Plant Pathology (3rd. edition). Academic Press, New York.

Åkerstrand K., Kardell L., Møller T. (1988) Investigation of blueberries and lingonberries from nitrogen fertilized plots (in Swedish). Vår Føda, 40, 259-70.

Alexandratos N. (Ed) (1988) World agriculture towards 2000. An FAO study. Belhaven Press, London.

Andersen J.M. (1986) Reversal of the eutrophication of the Baltic Sea. Vatten 42, 36-40.

Aniansson B. (1989) Northern Europe's seas - Northern Europe's environment. Nordic Council, Stockholm.

Anon. (1989a) Cadmium in phosphates: one part of a wider environmental problem. Phosphorus and Potassium no. 162, 23-30.

Anon. (1989b) Week 40 in Stockholm supermarkets: alternative produce for near 4 SEK more a day (in Swedish). Alternativodlaren (10) 15.

Arden-Clarke C., Hodges R.D. (1987) The environmental effects of conventional and organic/biological farming systems. I. Soil erosion, with special reference to Britain. Biol. Agric. Hort. 4, 309-57.

Arden-Clarke C., Hodges R.D. (1988) The environmental effects of conventional and organic/biological farming systems. II. Soil ecology, soil fertility and nutrient cycles. Biol. Agric. Hort.. 5, 223-87.

Aslyng H.C. (1978) Environment and agriculture (in Danish). DSR, Copenhagen.

Aslyng H.C. (1988) Phosphorus and nitrogen in soil and water (in Danish). Vand og Miljø 5 (2) 47-50.

Augstburger F., Zemp J., Heusser H. (1989) Vergleich der Fruchtbarkeit, Gesundheit und Leistung von Milchkühen in biologisch und konventionell bewirtschafteten Betrieben. Landwirt. Schweitz 1 (7) 427-31.

Bærug R., Singh B.R. (1987) The influence of long term use of fertilizers on the content of Cd in soils and plants. In: Lindberg S.E. Hutchinson T.C. (Eds) Heavy metals in the environment. Internat. Conf., New Orleans, Vol. 1, pp. 439-41. CEP Consultants, Edinburgh.

Bærug R., Singh B.R., Selmer-Olsen A.R., Håland A., Myhr K., Steinnes E. (1989) Effect of phosphorus fertilization on Cd content of soils and plants from southern and central parts of Norway. In: Vernet J-P (Ed) Heavy metals in the environment. Internat. Conf. Geneva, Vol. 2, pp. 32-35. CEP Consultants, Edinburgh.

Barber S.A. (1984) Soil nutrient bioavailability. A mechanistic approach. J. Wiley & Sons, New York.

Barnes R.S. Mann K.H. (Ed) (1980) Fundamentals of aquatic ecosystems. Blackwell Sci. Publ. Oxford.

Barraclough P.B. (1989) Root growth and nutrient uptake by field crops under temperate conditions. Aspects of Appl. Biol. 22, 227-33.

BASF (1985) Unser Boden. 70 Jahre Agrarforschung der BASF AG. Verlag Wissenschaft und Politik. Bibliothek Technik und Gesellschaft. Ludwigshafen..

Blum W.E.H., Herbinger B., Mentler A., Ottner F., Pollak M., Unger E., Wenzel W.W. (1989) Zur Verwendung von Gesteinsmehlen in der Landwirtschaft. I. Chemisch-mineralogische Zusammensetzung und Eignung von Gesteinsmehlen als Düngemittel. II. Wirkung von Gesteinsmehlen als Bodenverbesserungsmittel. Z. Pflanzenernähr. Bodenk., 152, 421-25, 427-30.

Böckenhoff E., Hamm U., Umhau M. (1986) Analyse der Betriebs und Productionsstrukturen sowie der Naturalertrage in Alternative Landbau. Ber. Landwirt. 64 (1) 1-39.

Bøckman O.C., Bryson D.D. (1989) Well-water methaemoglobinaemia: the bacterial factor. In: Wheeler D., Richardson M.L., Bridges J. (Eds) Watershed 89. The future for water quality for Europe. Vol. II pp. 239-44. Pergamon Press, Oxford.

Boeringa R. (Ed.) (1980) Alternative methods of agriculture. Agriculture and Environment 5 (special issue). (A shortened translation of the Dutch report (1976): Alternative landbouw methoden. Wageningen).

Boeringa R., de Vries A. (1989) Consumer quality of produce. In. Zadoks J.C.(Ed) Development of farming systems. Pudoc, Wageningen.

Bouwman A.F. (Ed) (1990) Soils and the greenhouse effect. J. Wiley & Sons, London.

Bowen H.J.M. (1979) Environmental chemistry of the elements. Academic Press, New York.

BP (1989) Statistical review of world energy. London.

Breirem K., Reisegg F., Höyem T., Njøs A., Rydlang K., Sande H., Wilhelmsen G. (1984) Energy use and food production in agriculture (in Norwegian). Agricultural Research Council of Norway. Landbruksforlaget, Oslo.

Brüne H. (1988) 100 Jahre VDLUFA: Bodenfruchtbarkeit - Rückblick und Ausblick. VDLUFA-Congress Bonn, VDLUFA - Schriftenreihe 27/1988, 165-7. Darmstadt.

Buijsman E., Maas H.F.M., Asman W.A.H. (1986) Antropogenic ammonia emissions in Europe. IMOU-report R-86-17 Utrecht, the Netherlands. Also in Atmos. Environ. (1987), 21(5) 1009-22.

Bulling M. (1986/7) Qualitätsvergleich von "Biologisch" und "Konventionell" erzeugten Feldfrüchten. Regierungspräsidium Stuttgart.

Bumb B. (1989) Global fertilizer perspective, 1960-95. The dynamics of growth and structural change. International Fertilizer Development Center, Muscle Shoals, Alabama.

Burrell A., Hill B.,Medland J. (1984) Statistical handbook of UK agriculture. Macmillan, London.

Burrell A., Hill B., Medland J. (1987) Statistical handbook of UK agriculture. Dept. of Agricultural Economics, Wye College, London

Campbell R., Macdonald R.M. (Eds) (1989) Microbial inoculation of crop plants. Sos. Gen. Microbiol. Special publ. 25. IRL Press at Oxford Univ. Press, Oxford.

CAST (Council for Agricultural Science and Technology) (1980) Organic and conventional farming compared. Report no. 84. Ames, Iowa.

CAST (Council for Agricultural Science and Technology) (1982). Soil erosion, its agricultural, environmental and socioeconomic implications. Report no. 92. Ames, Iowa.

CEA, IFA, IPI (1983) Handbook on environmental aspects of fertilizer use. Martinus Nijhoff/ Dr W. Junk Publishers, the Hague.

Clark C.W. (Ed) (1989) Managing planet earth. Scientific American 261 (3) 19-120.

Clark F.E., Rosswall T. (Eds) (1981). Terrestrial nitrogen cycles. Processes, ecosystem strategies and management impacts. Ecol. Bull. no. 33. Swedish Natural Sciences Research Council, Stockholm.

Clark R.B. (1989) Marine pollution (2 ed). Clarendon Press, Oxford.

Commission of European Communities (1978) Agricultural structure 1950 - 1976. Eurostat, Luxembourg/Bruxelles.

Commission of European Communities (1985a) Farm structure survey. Eurostat, Luxembourg/Bruxelles.

Commission of European Communities (1985b) Yearbook of agricultural statistics. Eurostat, Luxembourg/Bruxelles.

Commission of European Communities (1988) Agriculture statistical yearbook 1988. Eurostat, Luxembourg/Bruxelles.

Conford P. (1988) The organic tradition. An anthology of writings on organic farming 1900-1950. Green books, Bideford, UK.

Cooke B.W. (1982) Fertilizing for maximum yield (3rd ed.). Collins, London.

Cooke G.W., Williams R.J.B. (1970) Losses of nitrogen and phosphorus from agricultural land. Water Treatmt. Exam. 19, 253-76.

Corré W.J., Breimer T. (1979) Nitrate and nitrite in vegetables. Litterature survey no. 39. Pudoc, Wageningen.

Council of Europe (1984) Pesticides (6th ed., new edition in print). Strasbourg.

Court M.N., Stephen R.C., Waid J.S. (1964) Toxicity as a cause of the inefficiency of urea as a fertilizer. J. Soil Sci. 15, 42-48.

Curl E.A., Truelove B. (1986) The rhizosphere. Springer-Verlag, Berlin/-Heidelberg.

Danish Agricultural Information Office (1989) Handbook of agronomy planning 1989/90 (in Danish). Århus.

Danish Government National Committee on Crop Husbandry (1985) Nitrate in water (in Danish). Århus.

Danish Government Plant Protection Centre (1986) Pesticides use, advantages, disadvantages, perspectives. A status report (in Danish). Danish Research Service for Soil and Plant Sciences. Report S 1820. Copenhagen.

Danish National Agency of Environmental Protection (1984) Quality criteria for certain substances in drinking water (in Danish). Copenhagen.

Danish National Food Agency (1989) Pesticide residues in Danish food. Publication no. 177. Copenhagen.

Davies G.P., Barraclough D. (1990) Nitrate leaching at Rushall farm, Wiltshire 1985-8. Soil Use Manag. (in press).

Diaper J., Lawrenson O. (1986) Nitrate content of vegetables commercially grown on land subject to inorganic or organic fertilizer. Env. Geochem. Health 8 (4) 105-6.

Dilz K. (1988) Efficiency of uptake and utilization of fertilizer nitrogen by plants. In: Jenkinson D.S., Smith K.A. (Eds) Nitrogen efficiency in agricultural soils, pp. 1-26. Elsevier, London.

Dixon R.O.D., Wheeler C.T. (1986) Nitrogen fixation in plants. Blackie, Glasgow.

DLG (1988) Bodenleben, Bodenfruchtbarheit, Bodenschutz. Deutsche Landwirtschafts Gesellschaft, Arbeiten der DLG, Band 191. Frankfurt aM.

EC (1982) Directive on the major accident hazards of certain industrial activities. 82/501/EC (O.J. L 230. 5.8.82). Amended 87/216/EC (O.J. L 85.28.3.87).

EC (1986) Directive on the protection of the environment and in particular the soil when sewage sludge is used in agriculture. 86/278/EC (O.J. L 181. 4. 7. 86).

ECETOC (1988) Nitrate in drinking water. European Chemical Industry Ecology and Toxicology Centre. Technical report No. 27. Brussels.

Eliassen A., Hov Ø., Iversen T., Saltbones J., Simpson D. (1988) Estimates of airborne transboundary transport of sulphur and nitrogen over Europe. EMEP / MSC-W Report 1188. Norwegian Meteorological Institute, Oslo.

Engelhard A.W. (Ed) (1989) Soilborne plant pathogens: management of diseases with macro and micronutrients. American Phytopathological Society APS-Press St. Paul, Minnesota.

FAO (1950 - 1958) Yearbook of food and agricultural statistics. Rome.

FAO (1959 - 1986) Production yearbook. Rome.

FAO (1960-1986) Fertilizer review. Rome.

FAO (1984) Fertilizer and plant nutrition guide. FAO fertilizer and plant nutrition bulletin 9. Rome.

FAO (1986) Committee on Forest Development in the Tropics: tropical forestry action plan. Rome. Also in: Unasylva (1986) 38 (152) 37-64.

FAO (1987a) Fertilizer strategies. Land and water development series no. 10. Rome.

FAO (1987b) The fifth world food supply survey. Rome.

FAO (1989) Sustainable agricultural production: implications for international agricultural research. Research and technology paper 4. Rome.

Follett R.F., Stewart J.W.B. Cole C.V. (1987) Soil fertility and organic matter as critical components of production systems. Soils Sci. Soc. Am. Special publication no. 19. Am. Soc. Agron. Publ. Madison, Wisconsin.

Forman D. (1989) Are nitrates a significant risk factor in human cancer? Cancer Surveys 8(2) 443-58.

Foth H.D., Ellis B.G. (1988) Soil fertility. John Wiley & Sons, New York.

Furrer O.J., Stauffer W. (1984) Einfluss von Bodennutzung und Düngung auf die Nitratauswaschung im Schweizerischem Mittelland. 96 VDLUFA-Kongress Karlsruhe, VDLUFA-Schriftenreihe 10, 48-9. Darmstadt.

Geissler T., Geyer B. (1980) Die Wirkung einer Stickstoffdüngung auf die Qualität von Feldgemüse. Arch. Gartenbau 28(4), 199-207.

Gerlach S.A. (1987) Nutrients - an overview. Int. Conf. on Environmental Protection of the North Sea. London 24-27 March 1987. Session two: nutrients.

Gesslein S., Jansson S.L. (1987) Results from the experiments with different cultivating systems at Bjärröd (in Swedish). Skånskt Lantbruk, 20 (7/8) 375-82.

Gourdine S.P., Traiger W.W., Cohen D.S. (1983) Health food stores investigations. J. Am. Dietic Assn. 83(3), 285-90.

Greenwood D.J., Cleaver T.J., Turner M.K., Hunt J., Niendorf K.B., Loquens S.M.H. (1980) I. Comparison of the effects of potassium fertilizer on the yield, potassium content and quality of 22 different vegetable and agricultural crops. II. Comparison of the effects of phosphate fertilizer on the yield, phosphate content and quality of 22 different vegetable and agricultural crops. III. Comparison of the effects of nitrogen fertilizer on the yield., nitrogen content and quality of 21 different vegetable and agricultural crops. J. Agric. Sci. Camb. 95, 441-56, 457-69, 471-85.

Greenwood D.J., Hunt J. (1986) Effect of nitrogen fertilizer on the nitrate contents of field vegetables grown in Britain. J. Sci. Food Agric. 37, 373-83.

Grierson D., Covey S.N. (1984) Plant molecular biology. Blackie & Son Ltd., Glasgow.

Gunnarsson O. (1983) Heavy metals in fertilizers, do they cause environmental and health problems? Fertilizers Agric. 37 no. 85, 27 - 42.

Gushee D.E. (1989) Global climate change. Chemtec 19 (8), 470-9.

Gysi C., Ryser J-P., Lüthi J. (1985) Auswertung von Nitratuntersuchungen am Kopfsalat in schweizerischen Gemüsebau. Schweiz. Landw. Fo. 24, (3/4) 203-214.

Hansen H. (1981) Comparison of chemical composition and taste of biodynamically and conventionally grown vegetables. Qual. plant. - Plant Foods Hum. Nutr. 30, 203-11.

Hansen J.A., Tjell J.C. (1981) Use of sludge in agriculture (in Danish). Polyteknisk Forlag, Lyngby.

Hansson A.C., Pettersson R., Paustian K. (1987) Shoot and root production and nitrogen uptake in barley, with and without nitrogen fertilization. J. Agron. Crop Sci. 158, 163-71.

Hauch R.D. (Ed) (1984) Nitrogen in crop production. Am. Soc. Agron. Madison, Wisconsin

Heichel G.H., Barnes D.K. (1984) Opportunities for meeting crop nitrogen needs from symbiotic nitrogen fixation. In: Kraal D.M. (Ed) Organic farming: current technology and its role in sustainable agriculture. Am. Soc. Agron. Special Publ. 46, pp 49-59. Madison, Wisconsin.

Helsel Z.R. (Ed) (1987) Energy in plant nutrition and pest control. Energy in world agriculture vol. 2. Elsevier, Amsterdam.

Henze J., Hansen H. (1988) Lagerräume für Obst und Gemüse. Kuratorium für Technik und Bauwesen in der Landwirtschaft e V. KTBL-Schrift 327. Darmstadt

Hileman B. (1990) Alternative agriculture. Chem. Eng. News 68(10) 26 - 40.

HMSO (1948 - 1988) Annual review of agriculture. London.

Høg K. (1985a) Cultivation methods influence on yield, quality, soil and environment. I. Fertilization (in Danish) Danish Research Service for Soil and Plant Sciences. Report S 1795. Copenhagen.

Høg K. (1985b) Cultivation methods influence on yield, plant quality, soil and environment. IV Alternative Agriculture (in Danish). Danish Research Service for Soil and Plant Sciences. Report S 1805. Copenhagen.

Hognestad P.T. (1987) Assessments of the environmental conditions in the Skagerrak and Kattegat. ICES (International Council for the Exploration of the Sea) . Coop. research report 149. Copenhagen

House of Lords (1989) Select Committee on the European Communities Report: Nitrate in water. Session 88-89. 16th report. HMSO, London.

Howarth R.W. (1988) Nutrient limitation of net primary production in marine ecosystems. Ann. Rev. Ecol. 19, 89-110.

Huber D.M., Arny D.C. (1985) Interaction of potassium with plant disease. In: Munson R.D. (Ed). Potassium in agriculture. Amer. Soc. Agron. Madison, Wisconsin.

IAEA (1984) Soil and fertilizer nitrogen. International Atomic Energy Authority, Techn. Reports Series no. 244. Vienna.

IFA (1986) The fertilizer industry - the key to world food supplies. International Fertilizer Association, Paris.

Jaep A. (1986) Konventionelle und alternative Lanbau - ein betriebswirtschaftlichen Vergleich. Ber. Landwirt. 64 (1) 40 - 73

James L.S. (1988) Principles of farm irrigation system design. J. Wiley & Sons, London.

James W.P.T., Ferro-Luzzi A., Isaksson B., Szostak W.B. (1988) Healthy nutrition. Preventing nutrition-related diseases in Europe. WHO, Copenhagen.

Jansson S.L. (1986) Soil biology - plant production - fertility studies. The crop residues as a component of soil fertility. J. Royal Swed. Acad. Agr. Forestry, Suppl. 18, 9-31.

Jensen B.E., Leth T. (1985) Healthfood stores stock. Report 2. Laboratory analyses and conclusions (in Danish). Department of the Environment. State Food Institute, Publ. 111. Copenhagen.

Jenkinson D.S. (1982) The nitrogen cycle in long-term field experiments. Phil. Trans. Royal Soc. Lond. B 296, 563-571.

Jenkinson D. (1988/89) The long-term effects of nitrogen fertilizers. The Farmers Club J. no. 96, Dec.-Jan., 24-30.

Johnston A.E. (1982) The effects of farming system on the amount of soil organic matter and its effect on yield at Rothamsted and Woburn. In: Boels D., Davies D., Johnston A.E. (Eds) Soil degradation, pp. 187-202. A.A. Balkema, Rotterdam.

Johnston A.E. (1989) Phosphorus cycling in intensive arable agriculture. In: Thiessen H (Ed) Phosphorus cycles in terrestrial and aquatic ecosystems. SCOPE-UNEP Regional Workshop 1. Europe, pp. 123-136. Proceedings. Saskatchewan Inst. Pedology. University of Saskatchewan, Saskatoon.

Jollans J.L. (1985) Fertilizers in UK farming. Centre for Agricultural Strategy Report 9. University of Reading, Reading.

Jones K.E., Johnston A.E. (1989) Cadmium in cereal grain and herbage from long-term experimental plots at Rothamsted, UK. Environ. Poll. 57 (3) 199 - 216.

Jürgens-Gschwind S., Altbrod J. (1981) Die Energiesituation der deutschen Landwirtschaft. BASF Agricultural Bull. 1/81.

Jürgens-Gschwind S. (1989) Ground water nitrates in other developed countries (Europe) - relationship to land use patterns. In: Follett R.F. (Ed) Nitrogen management and ground water protection, pp. 75-138. Elsevier Sci. Publ. Amsterdam.

Kahnt G. (1986) Biologischer Pflanzenbau. Möglichkeiten und Grenzen biologischer Anbausysteme. Verlag Eugen Ulma, Stuttgart.

Khasawneh F.E., Doll E.C. (1978) The use of phosphate rock for direct application to soils. Adv. Agron. 30, 159-206.

Kjellerup V. (1983) Nitrogen fertilization influence on the content of nitrate-nitrogen in draining water 1973-81 (in Danish). Danish Research Service for Soil and Plant Sciences. Communication 1736. Copenhagen.

Knöppler H-O., Averdunk G. (1986) Vergleichende Qualitätsuntersuchungen von konventionell und alternativ erzeugter Kuhmilch. Arch. Lebensmittelhyg. 37 (4) 94-6.

Kofoed A. Dam, (Ed) (1985) Animal manure and its use (in Danish). Danish Research Service for Soil and Plant Sciences. Report S 1809, Copenhagen.

Krajnc E.I., van Gestel C.A.M., Mulder H.C.M., de Vrijer Fl., Sinkeldam E.J., Vink G.J., Canton J.H., van Apeldoorn M.E., Janus J.A. (1987) Integrated criteria document cadmium. National Institute of Public Health and Environmental Protection. Appendix to Report no. 758476002. Bilthoven, the Netherlands.

Kuhlmann H., Engels T. (1989) Nitrogen utilization in relation to N-fertilization. The Fertilizer Society, London, Proc. no. 287.

Kuipers H. (1982) Processes in physical soil degradation in mechanized agriculture. In: Boels D., Davies D., Johnston A.E. (Eds). Soil degradation, pp. 7-18. A.A. Balkema, Rotterdam.

Larsson K. (1988) Nitrogen and phosphorus as biomass limiting factors in the sea (in Swedish). Vatten, 44, 19-28.

Lee K.E. (1985) Earthworms. Their ecology and relationships with soil and land use. Academic Press, Sydney.

Lidgate H.J. (1987) Nutrient in the North Sea - A fertilizer industry view. Int. Conf. on Environmental Protection of the North Sea. London, 24-27 March 1987. Session two: nutrients.

Lieberman M., Coursey D.G. (Eds) (1983) Post harvest losses in perishable foods in the developing world. Post harvest physiology and crop preservation. Plenum Press, New York.

Loue A. (1986) Les oligo-éléments en agriculture. Agri-Nathan Inst. Paris.

Macdonald A.J., Powlson D.S., Poulton P.R., Jenkinson D.S. (1989) Unused fertilizer nitrogen in arable soils - its contribution to nitrate leaching. J. Sci. Food Agric. 46, 407-19.

Machet J-M. (1987) L'apport des techniques isotopiques à la stratégie de fertilisation. Perspectives Agricoles no. 115 (Juin) 37-42.

Machet J-M., Pierre D., Recours S., Remy J-C. (1987) Signification du coéfficient d'utilisation et conséquences pour la fertilisation azotée des cultures. C.R. Acad. Agric. Fr. 73 (3) 39-55.

MAFF (1989) Report of the working party on pesticide residues: 1985-88. UK. Ministry of Agriculture, Fisheries and Food. Food surveillance paper no. 25. HMSO, London.

Marschner H. (1986) Mineral nutrition of higher plants. Academic Press, London.

Marschner H., Rømheld V., Horst W.J., Martin P. (1986) Root induced changes in the rhizosphere: importance for the mineral nutrition of plants. Z. Pflanzenernähr. Bodenk. 149, 441-56.

Medawar P., Medawar J. (1985) Aristotle to Zoos. A philosophical dictionary of biology. Oxford Univ. Press, Oxford.

Mengel K. (1979) Influence of exogenous factors on the quality and chemical composition of vegetables. Acta Hort. 93, 133-51.

Mengel K., Kirkby E.A. (1987) Principles of plant nutrition (4th ed.). Int. Potash Institute, Bern.

Mertz (Ed) (1986) Trace elements in human and animal nutrition (5th ed.). Vol. 1 and 2 (1987). Academic Press, London.

Mudahar M.S., Hignett T.P. (1982) Energy and fertilizer. Policy implications and options for developing countries. IFDC (International Fertilizer Development Center) Tech. Bull. T-20. Muscle Shoals, Alabama.

Newbould P. (1982) Losses and accumulation of organic matter in soils. In: Boels D., Davies D., Johnston A.E. (Eds) Soil degradation, pp. 107-131. A. Balkema, Rotterdam.

Nielsen N.E. (1983) Plant parameters controlling the efficiency of nutrient uptake from the soil. In: UN Econ. Comm. Europe. Efficient use of fertilizers in agriculture. Dev. Plant Soil Sci. 10, pp. 199-219. Kluwer Academic Publ. (Martinius Nijhoff Publ.), the Hague.

Nilson T. (1979) Yield, storage ability, quality and chemical composition of carrot, cabbage and leek at conventional and organic fertilizing. Swedish Univ. Agric. Sci, Dept of Hort. Sci. Report 7, Uppsala. Acta Hort. 93, 209-23.

Nilsson L.G. (1986) Data of yield and soil analyses in long-term soil fertility experiments. J. Royal Swedish Acad. Agr. Forestry. Suppl. 18, 32-70.

NLVF (1983) Alternative Agriculture (in Norwegian). NLVF-survey 127. The Agricultural Research Council of Norway, Oslo.

Norwegian National Nutrition Council (1989) The Norwegian diet and nutrition and food policy. Oslo.

Olsen R.A. (1987) The use of fertilizers and soil amendments. In:Wolman M.G., Fournier F.G.A. (Eds) Land Transformation in Agriculture, pp. 203-26. John Wiley & Sons Ltd., Chichester.

Olsen Y., Jensen A. (1989) Status for NTNF eutrophication research programme. (in Norwegian). Norwegian Technical Scientific Research Council. SINTEF, Trondheim.

Ormberg S. (1987) Radioactive elements in NPK fertilizers in the Nordic countries In: Låg J. (Ed) Commercial fertilizers and geomedical problems, pp. 143-53. Norwegian University Press, Oslo.

Ormberg S. (1988) Fertilizer consumption in Norway and the natural radioactivity in the raw materials. In: Låg J. (Ed) Health problems in connection with radiation from radioactive matter in fertilizers, soils and rocks, pp. 128-37. Norwegian University Press, Oslo.

Østergaard H.S. (1988) Nitrogen predictions and nitrogen losses (in Danish). In: Schriver K: Review of Danish field experiements, pp. 95-105. Government Committee on Crop Husbandry, Århus.

Owen T.R., Jürgens-Gschwind S. (1986) Nitrates in drinking water: a review. Fertilizer Res. 10, 3-25.

Padmos L. (1986) Nitrogen fertilization of potatoes: effect on yield and quality. Netherlands Fertilizer Institute, Techn. Bull. no 16. the Hague.

Paul E.A., Clark F.E. (1989) Soil microbiology and biochemistry. Academic Press, San Diego, California.

Pimentel D., Pimentel M. (1979) Food, energy and society. Edward Arnold, London.

Postgate J.R. (1982) The fundamentals of nitrogen fixation. Cambridge University Press, Cambridge.

Potash and Phosphate Institute (1989) Conventional and low input agriculture. Economic and environmental evaluation, comparisons and considerations. Report. Atlanta, Georgia.

Prins W.H., Ditz K., Neeteson J.J. (1988) Current recommendations for nitrogen fertilization within the EC in relation to nitrate leaching. The Fertilizer Society, London, Proc. no. 276.

Prummel J. (1981) Kalidüngung und Kartoffelqualität. Der Kartoffelbau 32 (3) 73-76.

Rennie J.R. (1985) Nitrogen fixation in agriculture in temperate regions. In: Evans H.J., Bottomley P.J., Newton W.E. (Eds) Nitrogen fixation research progress, pp. 659-65. Martinius Nijhoff Publ., the Hague.

Robson A.D. (Ed) (1989) Soil acidity and plant growth. Academic Press, Sydney.

Ros J.P.M., Slooff W. (Eds) (1988) Integrated criteria document cadmium. National Institute of Public Health and Environmental Protection. Report no. 758476002 (in Dutch), no. 758476004 (in English). Bilthoven.

Royal Society (1983) The Nitrogen Cycle of the United Kingdom. London.

Royal Swedish Academy for Forestry and Agriculture (1987) Future farming and forestry (in Swedish: För framtida bruk). LT Publishers, Stockholm.

Rushton S.P. (1988) Earthworms in pastoral agriculture. Outlook on Agric. 17 (2) 44-48.

Sakshaug E. (1988) Phosphorus and nitrogen as limiting factors for algeal growth (in Norwegian). Vatten 44, 29-32.

Samaras I. (1977) Nachernteverhalten Unterschiedlich Gedüngter Gemüsearten mit besonderer Berüchsichtigung Physiologisher und Mikrobiologischer Parameter. Diss. Justus-Liebig-Universität, Giessen.

Sauerbeck D.R. (1982) Influence of crop rotation, manurial treatment and soil tillage on the organic matter content of German soils. In: Boels D., Davies D., Johnston A.E. (Eds) Soil degradation, pp. 163-178. A. Balkema, Rotterdam.

Scheffer F., Schachtschabel P. (1976) Lehrbuch der Bodenkunde (9. ed.). F. Enkes Verlag, Stuttgart.

Schupan W. (1974) Nutritional value of crops as influenced by organic and inorganic fertilizer treatments. Qual. Plant - Plant Foods Hum. Nutr. 23 (4) 333-58.

Schwertmann U., Rickson R.J., Auerswald K. (Eds) (1989) Soil erosion protection measures in Europe. Proc. EC workshop, Freising FRG May 24-26, 1988. CATENA Verlag, Cremlingen-Destedt, Soil technology series 1.

Shen S.M., Hart P.B.S., Poulson D.S., Jenkinson D.S. (1989) The nitrogen cycle in the Broodbalk wheat experiment. N^{15} labelled fertilizer residues in the soil and in the soil microbial biomass. Soil Biol. Biochem. 21, 529-33.

Skøien S. (1988) Soil erosion and runoff losses of phosphorus, effect of tillage and plant cover (in Norwegian). Norw. Agric. Res. 2 (4) 207-18.

Söderlund R., Rosswall T. (1982) The Nitrogen Cycle. In Hutzinger O. (Ed) The handbook of environmental chemistry, Vol. 1 B. The natural environment and the biogeochemical cycles, pp. 60-81. Springer Verlag, Berlin/Heidelberg.

Söderström J. (1988) Coastal growth of plankton, a steady state system controlled by phosphorus supply (in Swedish). Vatten 44, 3-10.

Soltner D. (1988) a) Les grandes productions végétales (16. ed.) b) Alimentation des animaux domestiques (18. ed.). Collection Sciences et Techniques Agricoles, Angers, France.

Spiertz J.H.H., de Vos N.M. (1983) Agronomic and physiological aspects of the role of nitrogen in yield formation of cereals. Plant Soil 75, 379-91.

Sprent J.I. (1988) The ecology of the nitrogen cycle. (Cambridge studies in ecology). Cambridge University Press, Cambridge.

Stapel C. (1982) Ecologic agriculture in global and national context (in Danish). Ugeskrift for Landbrug 127 (26) 495-500.

Stopes C., Woodward L., Forde G., Vogtman H. (1988) The nitrate content of vegetable and salad crops offered to the consumer as from "Organic" or "Conventional" production systems. Biol. Agric. Hortic. 5, 215-221.

Strebel O., Duynisveld W.H.M., Böltcher J. (1989) Nitrate pollution of ground water in Western Europe. Agric. Ecosystems Environ. 26, 189-214.

Syers J.K., Springett J.A. (1984) Earthworms and soil fertility. Plant Soil 76, 93-104.

Tisdale S.L., Nelson W.L., Beaton J.D. (1985) Soil fertility and fertilizers (4th ed.). Macmillan Publishing Co., New York.

Tudge C. (1988) Food crops for the future. Basil Blackwell Ltd., Oxford.

U.K. Department of the Environment (1987) Second international conference on the protection of the North Sea. Quality status of the North Sea. A report by the scientific and technical working group. HMSO, London.

UK Stratospheric Ozone Review Group (1987) Stratospheric ozone. HMSO, London.

UN (1986) Department of International Economic and Social Affairs: World population prospects: estimates and projections as assessed in 1984, Population Studies No. 98. New York.

UN (1989) Prospects of world urbanization, 1988. Dept. of Int. Econ. and Social Affairs. Population studies no. 112. New York.

US Bureau of Mines (1985) Mineral facts and problems. Bull. 675. Washington D.C.

US National Research Council (1989) Alternative Agriculture. Washington D.C.

van der Meer H.G., Unwin R.J., van Dijk T.A., Ennik G.S. (Eds.) (1987) Animal manure on grassland and fodder crops. Fertilizer or waste? Martinius Nijhoff Publ. Dordrecht.

van der Meer H.G., Baan Hofman T. (1989) Contributions of legumes to yield and nitrogen economy of lays on a biodynamic farm. In: Plancquaert P., Hagga R. (Eds) Legumes in farming systems, pp. 25-36.. Kluwer Acad. Publ. Doordrecht.

van Duijvenbooden W., Matthijsen A.J.C.M. (Eds) (1989) Integrated criteria document nitrate. Dutch National Institute of Public Health and Environmental Protection. Report no. 758473012. Bilthoven.

van Stavern W.A., Dhuyvetter H.M., Bons A., Zeelen M., Hautvast J.G.A. (1985) Food consumption and height/weight status of Dutch preshool children on alternative diets. J. Am. Dietic Assn. 85, 1579-84.

Vereijken P. (1986) From conventional to integrated agriculture. Netherlands J. Agric. Sci. 34, 387-93.

Vereijken P. (1989) Experimental systems of integrated and organic wheat production. Agr. Systems. 30, 187 - 97.

Vetter H. (1980) Umwelt und Nahrungsqualität. Wm. Heyne Verlag, München.

Vetter H., Kampe W., Ranfft K. (1983) Qualität pflanzlicher Nahrungsmittel. Verband Deutsche Landwirtschaftliche Untersuchungs und Forschungsanstalten. VDLUFA Schriftenreihe heft 7, Darmstadt.

Vogtmann H., Temperli A.T., Künsch U., Eichenberger M., Ott P. (1984) Accumulation of nitrate in leafy vegetables grown under contrasting agricultural systems. Biol. Agric. Hort. 2, 51-68.

Warneck P. (1988) Chemistry of the natural atmosphere. Academic Press, San Diego, California.

Webster C.C., Wilson P.N. (1980) Agriculture in the tropics (2 ed.). Longman, London.

Wetselaar R., Farquhar G.D. (1980) Nitrogen losses from tops of plants. Adv. Agron. 33, 263-302.

WHO (1985a) Energy and protein requirements. Technical reports series 724. Geneva.

WHO (1985b) Health hazards from nitrates in drinking-water. Report on a WHO meeting 5-9 March 1984. Copenhagen.

Wiedemann-Sander A. (1987) Düngung im Zwiespalt. Fluch oder Segen? Verlagsges. für Ackerbau, Kassel.

Wild A. (1988) Russell's soil conditions and plant growth (11th ed.). Longman Scientific & Technical, Harlow.

Wilson J.R. (Ed) (1988) Advances in nitrogen cycling in agricultural ecosystems. CAB International, Wallingford, UK.

Wolman M.G., Fournier F.G.A. (Eds.) (1987) SCOPE 32: Land transformation in agriculture. John Wiley & Sons, London.

Wookey B. (1987) Rushall. The story of an organic farm. Basil Blackwell, Oxford.

World Bank (1986) World development report, 1986. Oxford University Press, Oxford.

World Commission (1987) World Commission on Environment and Development. Brundtland G. (Chairman): Our common future. Oxford University Press, Oxford.

World Resources Institute (1987) World Resources 1987. Basic Books Inc., New York.

Worldwatch Institute (1989) Brown L.R., (Ed) State of the world 1989. W.W. Norton & Co. New York.

Zadoks J.C. (Ed) (1989) Development of farming systems. Pudoc, Wageningen.

UNITS AND GLOSSARY

Units

Nitrogen is given as the element N.

Nitrate concentrations in water are given as nitrate, NO_3^-, to conform with EC practice.

Phosphate and potassium are given as the elements P and K. To convert to the oxides multiply:

$N \cdot 4.43 = NO_3^-$	$NO_3^- \cdot 0.23 = N$
$P \cdot 2.3 = P_2O_5$	$P_2O_5 \cdot 0.43 = P$
$K \cdot 1.2 = K_2O$	$K_2O \cdot 0.83 = K$

Area:
1ha = 10 da = 10 000 m^2 = 2.47 acres

Glossary

Agronomy	The branch of agriculture dealing with crop production.
Alternative agri-culture	A collective term for agricultural practices that reject the use of soluble mineral fertilizers and pesticides.
Ammonium nitrate	NH_4NO_3, made from nitric acid and ammonia. Common fertilizer, with 34 per cent nitrogen, one half as ammonium, one half as nitrate. Also used in manufacture of explosives.
Ammonium sulphate	$(NH_4)_2SO_4$, ammonium salt of sulphuric acid. Traditional fertilizer supplying both nitrogen and sulphur, first produced as by-product in coal-gas manufacture.
Anaerobic	Oxygen deficient conditions.
Anion	A negatively charged atom or group of atoms, e.g., nitrate (NO_3^-) or sulphate (SO_4^{2-}).
Application	General term covering all processes for giving fertilizers to crop or soil.
Aquifer	A layer of rock which holds water and allows water to percolate through it.

Arable	Farming based on annual ploughing or cultivation of the land and sowing of an annual crop.
Becquerel	Unit of radioactivity: 1 disintegration per second.
Calcareous	Containing calcium usually in the form of chalk or limestone. Normally a description of soils.
Calcination	Heating to high temperature.
Calcium ammonium nitrate	A mixture of ammonium nitrate and pulverized limestone or related material, made into particles.
Calcium nitrate	Calcium salt of nitric acid, a non-acidifying fertilizer with 15 per cent N.
Canopy	Above ground structure of a crop comprising leaves, stems and branches.
CAP	Abbreviation for Common Agricultural Policy of the European Community.
Catch crop	An extra crop, usually quick-growing, grown between two main crops in a rotation.
Cation	Positively charged atom or group of atoms, e.g., potassium (K^+), ammonium (NH_4^+).
Cereal	Cultivated member of the grass family whose seeds or grain are used for food or animal feed: wheat, barley, rye, oats, rice.
Clay	A constituent of soil comprising very fine particles less than 0.002 mm in diameter.
Club root	A fungal disease (*Plasmodiophora brassicae*) of cruciferous plants, particularly brassicae, e.g., swedes, rape, cabbage.
Compaction	Crushing and compression of the soil caused by vehicles or animal treading.
Complex fertilizer	Mineral fertilizer containing two or more of the major nutrients, N, P and K.
Contour ploughing	Ploughing on the level, following the contour as distinct from up and down the slope. A soil erosion control measure.

Conventional agriculture	Another name for current agriculture.
Cover-crop	A crop which provides protection to a second crop grown beneath it.
Crucifer	A plant belonging to the cabbage family.
Cultivation	A tillage operation on the land involving inversion or mixing of the surface layers of the soil.
Current agriculture	Agriculture conforming to good agricultural practices according to local conditions, including the use of fertilizers and pesticides.
Drilling	Sowing of seed in rows at the required depth in the soil. Fertilizer may be placed simultaneously with the seed.
EC9	The nine member states of the EC in 1973 - France, Federal Republic of Germany, Italy, Netherlands, Belgium, Luxembourg, UK, Denmark and Eire.
EC10	The ten member states in 1981 - nine as above plus Greece.
EC12	The twelve member states from 1986 - ten as above plus Spain and Portugal.
Enzymes	Proteins which act as a catalyst for chemical changes, e.g., urease catalyses the splitting of urea to ammonia and carbon dioxide.
Evapotranspiration	Loss of water from the land by evaporation from the soil surface and by transpiration from plant leaves.
Extension service	Agricultural advisory and development service for farmers.
Fallow	Land left unsown, usually for whole or part of a season during which it is ploughed or cultivated to kill weeds.
Fungicide	A chemical used to kill fungi and so control fungal disease, usually in crops.
GATT	Abbreviation for General Agreement on Trade and Tariffs.
Goitre	Enlargement of the human thyroid gland. One cause of this disorder is iodine deficiency in the diet.
Green manure	A crop specifically grown for subsequent ploughing-in.

Haemoglobin	The red pigment in blood cells which acts as a carrier for oxygen.
Half-life	The period in which the activity of a substance falls by 50%. Usually applied to radioactivity.
Harrowing	Light cultivation of land using an implement equipped with tines or chains.
Heavy metal	Metallic element with high specific weight, often toxic to mammals, e.g., cadmium, lead.
Herbicide	A chemical used to kill weeds.
Humus	Soil organic residue transformed to such a degree that it has lost its original structure. Some of it may decompose further over a span of a few years, other fractions may persist for centuries.
Igneous	Volcanic origin, usually applied to rocks.
Incidence	Level of occurrence of a disease or pest.
Inter-crop	Growing of two or more different crop species in the same field at the same time.
Isotope	Form of an element differing from other forms in the number of neutrons in the nucleus and so in atomic weight.
Legume	A plant of the pea family.
Ley	Land temporarily (from one up to ten years) sown to grass or grass-clover mixture.
Liming	Application to land of a material containing calcium, usually chalk or limestone, in order to reduce soil acidity.
Loamy	Soil with a balanced particle mixture, approximately 25 % clay, 40% sand and 35% silt.
Median	Point in a series of numbers at which 50% of the numbers are greater and 50% smaller.
Methane	Gaseous compound of carbon and hydrogen (CH_4) inflammable and produced during the anaerobic decomposition of organic material. Main constituent of natural gas.

Mineralisation	Conversion of soil organic matter through microbiological and chemical processes into inorganic crop nutrients.
Mite	Sub-class of arachnida allied to ticks, often parasitic on animals or crops.
Mixed farming	Maintenance of arable and livestock enterprises on the same farm.
Monoculture	Growing of the same crop on the same field year after year.
Mulch	Material such as straw, leaves, etc., applied to the soil surface to conserve moisture and suppress weeds.
Nematode	A class of roundworms, mainly animal or crop parasites but also free-living in soil or water.
Nitrophosphates	Fertilizers made from phosphate rock and nitric acid, alone or mixed with other acids, usually with ammonia added.
No-till	Arable cropping where soil cultivation is not practiced.
Optimum-Optimal	Combination of factors giving best result. In fertilizer recommendations usually used for the application rate to a crop which gives the greatest economic return.
Potassium chloride	Potassium fertilizer material, produced from natural deposits of the mineral.
Pathogen	An organism which causes disease.
Perennial	A plant or crop which continues growth from year to year.
Pesticide	A chemical used to kill pests in crops or animals, primarily insects; sometimes also includes fungicides and herbicides.
pH	A measure of hydrogen ion activity and so of acidity or alkalinity. pH 7 is neutral reaction in water. Values below 7 indicate acidity, above 7 alkalinity.
Photosynthesis	The process by which green plants synthesize carbohydrates from carbon dioxide and water.
Ploughing	Mechanical inversion of the topsoil.
Predator	An animal which kills other animals for food.

Rotavating	Cultivation of the soil surface using a machine equipped with a rotating horizontal shaft fitted with tines or blades.
Sand	A mineral constituent of soils with a particle size in the range 0.05 to 2.0 mm.
Silt	A constituent of soil comprising particles intermediate in size between clay and sand - 0.002 - 0.05 mm in diameter.
Slurry	Mixture of urine and dung, with or without dilution with water.
Specialised arable agriculture	Continuous growth of arable crops over a whole farm or large area of land.
Spring crops	Crops sown in the late winter or spring.
Stubble	The part of the crop left in the ground after harvest including stems and roots.
Sub-tropical	Areas just outside the tropics.
Superphosphate	Phosphate fertilizer made by treating phosphate rock with sulphuric acid to make a water soluble product. When phosphoric acid is used, the product is called tripple superphosphate.
Sustainable	Practice or process that meets the needs of the present without compromising the ability to meet those of the future (World Commission on Environment and Development).
Sward	The carpet of grasses or clovers covering the ground in a pasture or meadow.
Take-all	A soil-borne fungus disease of wheat and barley (*Gaumannomyces graminis*) which affects the root and stem base.
Tillage	Practice of cultivation of soil. Also used to describe arable land.
Tillering	Development of sideshoots from the base of a plant close to the ground, e.g., from the bottom of the stem of cereals.
Topsoil	Top layer of the soil, some 20 - 30 cm deep.

Undersowing	Sowing two crops in combination so that one (a low undercrop) can continue growing after the main (cover) crop is harvested. Usually a grass or grass-clover seed mixture with cereals.
Urea	$(NH_2)_2$ CO, the end product of nitrogen metabolism in mammals excreted with urine. Also produced industrially as a fertilizer material from ammonia and carbon dioxide.
Waterlogged	Soil saturated with water so that all pore space is completely filled with water.
Weathering	Processes by which rocks disintegrate, eventually producing soil particles.
Winter crops	Crops sown in the late summer, autumn or early winter.

LIST OF FIGURES

LIST OF TABLES

SUBJECT INDEX